10.99

AQA AS CHEMISTRY

Rob King
Editor: Graham Curtis

LRC
St. Francis Xavier College
Malwood Rd
London SW12 8EN

Hodder Education, an Hachette UK company, 338 Euston Road, London NW1 3BH

Orders
Bookpoint Ltd, 130 Milton Park, Abingdon, Oxfordshire OX14 4SB
tel: 01235 827827
fax: 01235 400401
e-mail: education@bookpoint.co.uk
Lines are open 9.00 a.m.–5.00 p.m., Monday to Saturday, with a 24-hour message answering service. You can also order through the Hodder Education website: www.hoddereducation.co.uk

© Rob King, Graham Curtis 2012
ISBN 978-1-4441-8092-3

First printed 2012
Impression number 5 4 3 2 1
Year 2017 2016 2015 2014 2013 2012

All rights reserved; no part of this publication may be reproduced, stored in a retrieval system, or transmitted, in any form or by any means, electronic, mechanical, photocopying, recording or otherwise without either the prior written permission of Hodder Education or a licence permitting restricted copying in the United Kingdom issued by the Copyright Licensing Agency Ltd, Saffron House, 6–10 Kirby Street, London EC1N 8TS.

Cover photo reproduced by permission of Alexey Shkitenkov/Fotolia

Typeset by Datapage (India) Pvt. Ltd.
Printed in India

Hachette UK's policy is to use papers that are natural, renewable and recyclable products and made from wood grown in sustainable forests. The logging and manufacturing processes are expected to conform to the environmental regulations of the country of origin.

P2181

Get the most from this book

Everyone has to decide his or her own revision strategy, but it is essential to review your work, learn it and test your understanding. These Revision Notes will help you to do that in a planned way, topic by topic. Use this book as the cornerstone of your revision and don't hesitate to write in it — personalise your notes and check your progress by ticking off each section as you revise.

☑ Tick to track your progress

Use the revision planner on pages 4 and 5 to plan your revision, topic by topic. Tick each box when you have:

- revised and understood a topic
- tested yourself
- practised the exam questions and gone online to check your answers and complete the quick quizzes

You can also keep track of your revision by ticking off each topic heading in the book. You may find it helpful to add your own notes as you work through each topic.

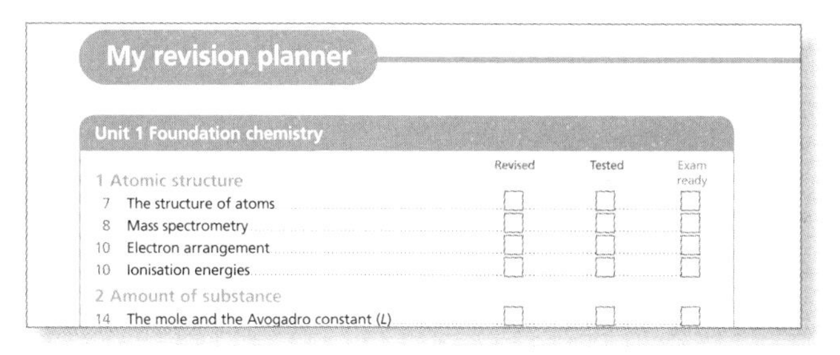

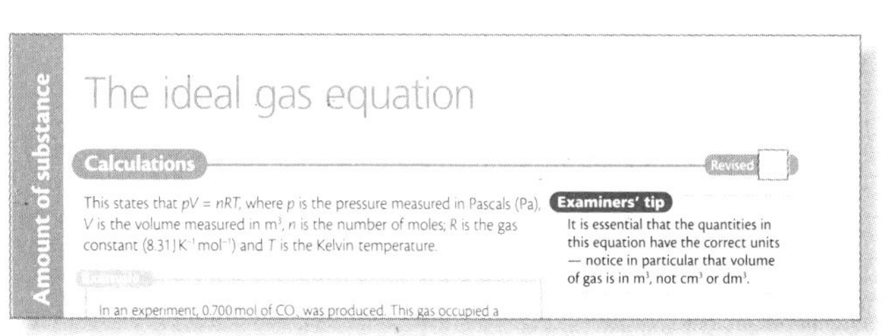

Features to help you succeed

Examiners' tips and summaries

Expert tips are given throughout the book to help you polish your exam technique in order to maximise your chances in the exam.

The summaries provide a quick-check bullet list for each topic.

Typical mistakes

The author identifies the typical mistakes candidates make and explain how you can avoid them.

Now test yourself

These short, knowledge-based questions provide the first step in testing your learning. Answers are at the back of the book.

Definitions and key words

Clear, concise definitions of essential key terms are provided where they first appear.

Key words from the specification are highlighted in bold throughout the book.

Revision activities

These activities will help you to understand each topic in an interactive way.

Exam practice

Practice exam questions are provided for each topic. Use them to consolidate your revision and practise your exam skills.

Online

Go online to check your answers to the exam questions and try out the extra quick quizzes at **www.therevisionbutton.co.uk/myrevisionnotes**

My revision planner

Unit 1 Foundation chemistry

Unit 2 Chemistry in action

Exam practice answers and quick quizzes at www.therevisionbutton.co.uk/myrevisionnotes

Countdown to my exams

6–8 weeks to go

- Start by looking at the specification — make sure you know exactly what material you need to revise and the style of the examination. Use the revision planner on pages 4 and 5 to familiarise yourself with the topics.
- Organise your notes, making sure you have covered everything on the specification. The revision planner will help you to group your notes into topics.
- Work out a realistic revision plan that will allow you time for relaxation. Set aside days and times for all the subjects that you need to study, and stick to your timetable.
- Set yourself sensible targets. Break your revision down into focused sessions of around 40 minutes, divided by breaks. These Revision Notes organise the basic facts into short, memorable sections to make revising easier.

Revised

4–6 weeks to go

- Read through the relevant sections of this book and refer to the examiners' tips, examiners' summaries, typical mistakes and key terms. Tick off the topics as you feel confident about them. Highlight those topics you find difficult and look at them again in detail.
- Test your understanding of each topic by working through the 'Now test yourself' questions in the book. Look up the answers at the back of the book.
- Make a note of any problem areas as you revise, and ask your teacher to go over these in class.
- Look at past papers. They are one of the best ways to revise and practise your exam skills. Write or prepare planned answers to the exam practice questions provided in this book. Check your answers online and try out the extra quick quizzes at **www.therevisionbutton.co.uk/myrevisionnotes**
- Use the revision activities to try out different revision methods. For example, you can make notes using mind maps, spider diagrams or flash cards.
- Track your progress using the revision planner and give yourself a reward when you have achieved your target.

Revised

One week to go

- Try to fit in at least one more timed practice of an entire past paper and seek feedback from your teacher, comparing your work closely with the mark scheme.
- Check the revision planner to make sure you haven't missed out any topics. Brush up on any areas of difficulty by talking them over with a friend or getting help from your teacher.
- Attend any revision classes put on by your teacher. Remember, he or she is an expert at preparing people for examinations.

Revised

The day before the examination

- Flick through these Revision Notes for useful reminders, for example the examiners' tips, examiners' summaries, typical mistakes and key terms.
- Check the time and place of your examination.
- Make sure you have everything you need — extra pens and pencils, tissues, a watch, bottled water, sweets.
- Allow some time to relax and have an early night to ensure you are fresh and alert for the examination.

Revised

My exams

AS Chemistry Unit 1

Date: ....

Time: ....

Location: ....

AS Chemistry Unit 2

Date: ....

Time: ....

Location: ....

1 Atomic structure

The structure of atoms

Protons, neutrons and electrons

Revised

Table 1.1 Fundamental particles

Particle	Relative charge	Relative mass
Proton	+1	1
Neutron	0	1
Electron	−1	$\frac{1}{1836}$

The **atomic number** (*Z*) and **mass number** (*A*) can be used to deduce the number of protons, neutrons and electrons in any atom or ion.

To calculate the number of neutrons in the nucleus, the atomic number (equal to the number of protons) is subtracted from the mass number (equal to the number of protons + neutrons).

For example, atoms of the elements calcium, fluorine and phosphorus are represented as

$^{40}_{20}Ca$ $^{39}_{19}K$ $^{31}_{15}P$

Atoms do not have an overall charge because the number of positively charged protons is the same as the number of negatively charged electrons. However, when an **ion** forms — either by losing or by gaining electrons — there is an overall charge because there will be an unequal number of positive charges and negative charges.

Atomic number is the number of protons in the nucleus of an atom (or ion).

Mass number is the total number of protons plus neutrons in the nucleus.

Examiners' tip

Remember that when atoms lose electrons they form positively charged ions; when atoms gain electrons they form negatively charged ions.

Isotopes

Revised

Many elements exist as **isotopes.** In a sample of chlorine, for example, 75% of the chlorine atoms have a mass number of 35 and 25% have a mass number of 37. These atoms have different masses because they have different numbers of neutrons.

In Table 1.2, the numbers of protons, neutrons and electrons are indicated for all three isotopes of carbon in which the differing numbers of neutrons are indicated.

Isotopes are atoms with the same number of protons, but different numbers of neutrons.

Table 1.2 The isotopes of carbon

Isotopes	Protons	Neutrons	Electrons
^{12}C	6	6	6
^{13}C	6	7	6
^{14}C	6	8	6

Mass spectrometry

The mass spectrometer

Revised

The **mass spectrometer** is a device that enables substances (elements or compounds) to be analysed by determining the masses of ions formed.

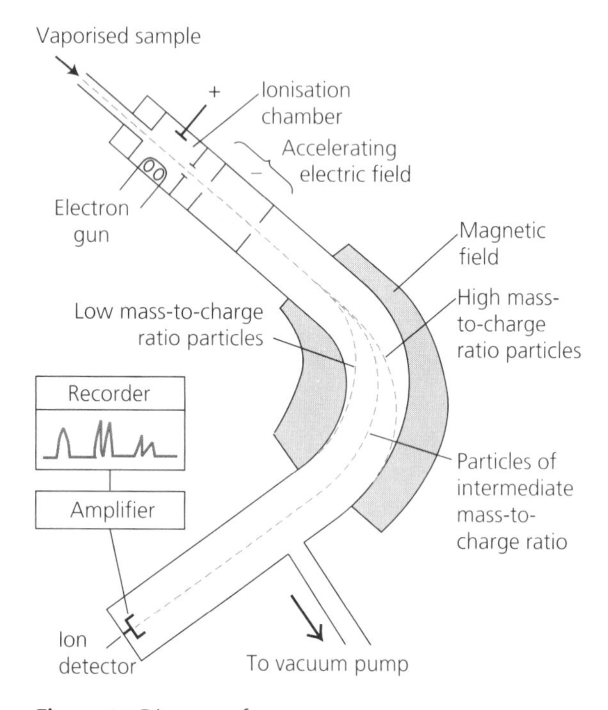

Figure 1.1 Diagram of a mass spectrometer

There are four stages involved in analysing a sample:

1. **Ionisation** — a high-energy beam of electrons from the electron gun removes the highest energy (outer) electron from a molecule or an atom, forming a positively charged ion. For example, using aluminium:
   $Al(g) \rightarrow Al^{+}(g) + e^{-}$
2. **Acceleration** — the positively charged ion is accelerated towards a negatively charged electrode; this provides all ions with the same kinetic energy.
3. **Deflection** — the positively charged ions are forced into a circular path using a strong magnetic field. Ions are sorted out according to their mass; the lower the mass of the ion the greater the degree of deflection.
4. **Detection** — ions of different masses are detected electronically. Each produces a small, electrical current that can be amplified and displayed on a computer, and a mass spectrum is produced.

Examiners' tip

Equations showing ionisation processes that take place in a mass spectrometer occur in the gas phase, so make sure that state symbols are added to equations showing ionisations.

Revision activity

Try to make up an acronym that will help you to remember the main processes involved when analysing a sample in a mass spectrometer.

How to interpret a mass spectrum

Revised

The mass spectrum of lead is shown in Figure 1.2.

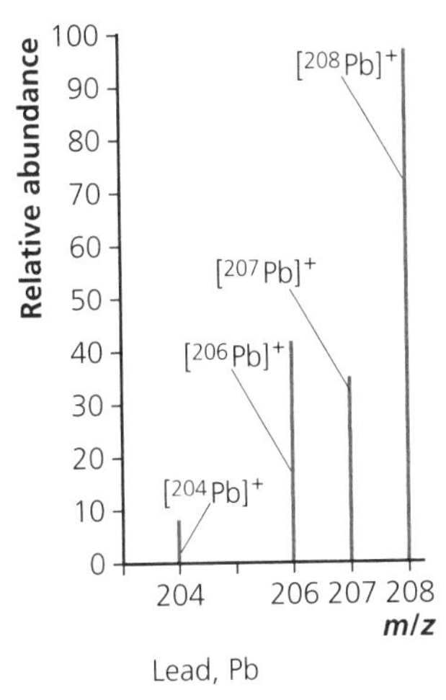

Figure 1.2 The mass spectrum of lead

- In the spectrum there are four peaks. This means that there are four isotopes of lead in the sample: ^{204}Pb, ^{206}Pb, ^{207}Pb and ^{208}Pb.
- There is more ^{208}Pb than any other lead isotope because this is the tallest peak — this isotope gives the peak with the highest abundance.
- You can calculate a value for the **relative atomic mass** (A_r) of lead by multiplying each mass number by its abundance (as a percentage) and then adding them all together. If the abundances in this case are ^{204}Pb (4.9%), ^{206}Pb (23.2%), ^{207}Pb (19.2%) and ^{208}Pb (52.7%), the calculation will be:

$$A_r \text{ for lead} = \left(\frac{4.9}{100} \times 204\right) + \left(\frac{23.2}{100} \times 206\right) + \left(\frac{19.2}{100} \times 207\right) + \left(\frac{52.7}{100} \times 208\right) = 207.1$$

(no units)

Mass spectrometers allow **relative molecular masses** to be determined with a high degree of precision. Mass spectrometers can be used in planetary space probes to analyse samples of material found on other planets, and this information can then be beamed to Earth.

Examiners' tip

In a mass spectrum, mass-to-charge ratio is measured on the horizontal axis and relative abundance is measured on the vertical axis. The charge of an ion is normally designed to be +1. In a few cases, a +2 ion may form when a second electron is removed.

Revision activity

Look up on the internet an element of your choice and find out about the relative abundances of its isotopes. Then work out the element's relative atomic mass for practice at this type of calculation.

Relative atomic mass (A_r) is the weighted mean mass of an atom compared to $\frac{1}{12}$ the mass of an atom of ^{12}C. ^{12}C has the value of 12.0000 on this scale.

Relative molecular mass (M_r) is the weighted mean mass of a molecule compared to $\frac{1}{12}$ the mass of an atom of ^{12}C. Relative molecular mass (M_r) is the sum of the relative atomic masses of the atoms in a molecule.

Now test yourself

Tested

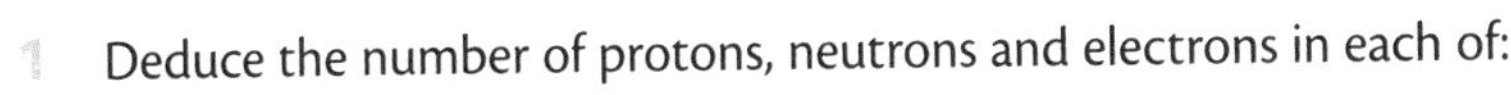

1. Deduce the number of protons, neutrons and electrons in each of:
   (a) $^{9}_{4}Be$
   (b) $^{31}_{15}P$
   (c) $^{24}_{12}Mg^{2+}$
   (d) $^{127}_{53}I^-$
2. Calculate the relative atomic mass for a sample of krypton from the data in the table below. Give your answer to 2 decimal places.

Isotopes	$^{78}_{36}Kr$	$^{80}_{36}Kr$	$^{82}_{36}Kr$	$^{83}_{36}Kr$	$^{84}_{36}Kr$	$^{86}_{36}Kr$
% abundance	0.35	2.3	11.6	11.5	56.9	17.4

3. A sample of boron is found to have a relative atomic mass of 10.8. Assuming that there are only two isotopes of this element, ^{10}B and ^{11}B, determine the percentage of each isotope in the sample.

Answers on p. 103

Electron arrangement

Electronic configurations

Revised

Electrons occupy **energy levels** (or shells) when orbiting an atomic nucleus. Energy levels are made up of **sub-levels**, or sub-shells.

Sub-levels fill up in the order shown in Figure 1.3, lowest energy first.

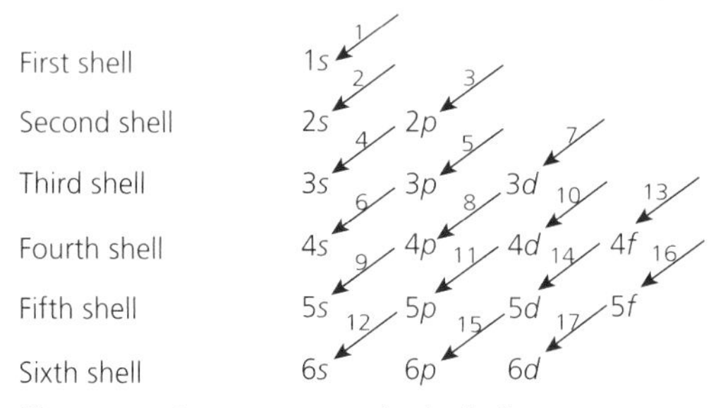

Figure 1.3 Arrangement of sub-shells

You can write an electronic configuration for an atom or an ion by putting the correct number of electrons into each sub-level: 2 electrons in an s-sub-level; 6 electrons in a p-sub-level; 10 electrons in a d-sub-level and 14 in an f-sub-level.

Table 1.4 Some electron configurations

Element	Symbol/ion	Atomic number	Electronic configuration
Sodium	Na	11	$1s^2, 2s^2, 2p^6, 3s^1$
	Na^+		$1s^2, 2s^2, 2p^6$
Potassium	K	19	$1s^2, 2s^2, 2p^6, 3s^2, 3p^6, 4s^1$
	K^+		$1s^2, 2s^2, 2p^6, 3s^2, 3p^6$
Sulfur	S	16	$1s^2, 2s^2, 2p^6, 3s^2, 3p^4$
	S^{2-}		$1s^2, 2s^2, 2p^6, 3s^2, 3p^6$
Gallium	Ga	31	$1s^2, 2s^2, 2p^6, 3s^2, 3p^6, 4s^2, 3d^{10}, 4p^1$
	Ga^{3+}		$1s^2, 2s^2, 2p^6, 3s^2, 3p^6, 3d^{10}$
Krypton	Kr	36	$1s^2, 2s^2, 2p^6, 3s^2, 3p^6, 4s^2, 3d^{10}, 4p^6$

Examiners' tip

Sub-levels are made up of **orbitals** and one orbital of any type can only hold up to two electrons.

Examiners' tip

Notice that the 4s-sub-level fills up before the 3d-sub-level; this means that the 4th energy level starts to fill up before the 3rd energy level has been filled completely.

Now test yourself

4 **(a)** Lithium (Z = 3), sodium (Z = 11) and potassium (Z = 19) are the first three members of group 1 of the periodic table. Write out each element's electronic configuration.

(b) An ion of charge +3 has the electronic configuration $1s^2$, $2s^2$, $2p^6$. What is the atomic number, Z, for the ion?

Answers on p. 103

Tested

Ionisation energies

First ionisation energy

Revised

The equation to represent the **first ionisation energy** of calcium is:

$$Ca(g) \rightarrow Ca^+(g) + e^-$$

A plot of first ionisation energy against atomic number is shown in Figure 1.4. This graph can provide valuable evidence for the electron arrangement in levels.

First ionisation energy is defined as the energy required to remove one mole of electrons from one mole of gaseous atoms.

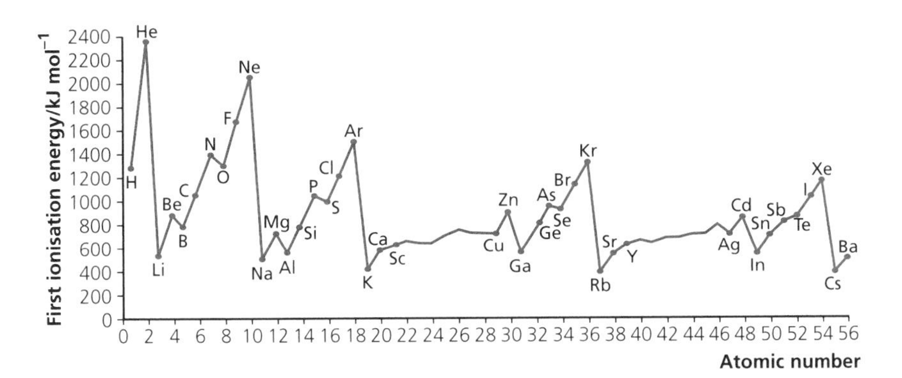

Figure 1.4 First ionisation energy graph

Energy levels

Revised

It is possible to remove more than one electron from an atom if the energy of the bombarding electrons is sufficient. For example, the 3rd ionisation energy of argon is represented as:

$$Ar^{2+}(g) \rightarrow Ar^{3+}(g) + e^-$$

Successive ionisation energies for elements give valuable evidence about the existence of energy levels.

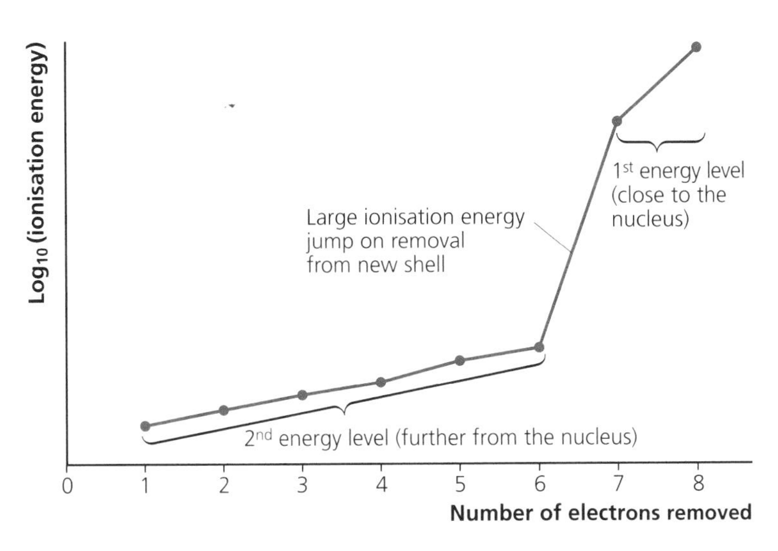

Figure 1.5 Graph of $\log_{10}$(ionisation energy) versus number of the electron being removed for oxygen

From Figure 1.5, it can be seen that:

- there is a large jump in ionisation energy required moving from the 6th to the 7th electron removal
- 6 electrons are relatively easy to remove from the nucleus, whereas 2 electrons are a lot more difficult to remove.

The explanation is that the 8 electrons must exist in two separate energy levels — 2 electrons are in an energy level closest to the positive charge of the nucleus and are therefore harder to remove. The other 6 electrons are in a higher energy level further from the nucleus and are therefore easier to remove.

Now test yourself

5 **(a)** Sketch a graph to show the successive ionisation energies for silicon, $Z = 14$.

(b) Write an equation to show the 4th ionisation energy of phosphorus, P.

Answers on p. 103

Evidence for energy levels

Variation of first ionisation energies down a group

The peaks in the graph of first ionisation energies (Figure 1.4) show that ionisation energy falls down group 0. This fall in ionisation energy applies on moving down all groups.

The explanation is:

- despite the increased nuclear charge, the electron being removed is in a new energy level which is progressively further from the nucleus
- the energy levels inside the extra energy level provide extra shielding for the removed electron from the attraction of the positively charged nucleus
- the net effect is to decrease the ionisation energy.

Evidence for energy sub-levels

Variation of first ionisation energies across period 3 from left to right

The trend in first ionisation energies is a general increase across period 3 from sodium to argon.

The explanation is

- the electrons are being removed from the same electron energy level
- the nuclear charge is increasing as more protons are being added from left to right
- the electron experiences a greater attraction as the atom increases in atomic number.

Although the general trend is for ionisation energy to increase from left to right across a period, there are two small decreases.

From group 2 to group 3: magnesium to aluminium

The explanation for the decrease is:

- despite the increased nuclear charge, the added electron is in a new p-*sub*-shell of slightly higher energy, and this is slightly further from the nucleus
- the s^2 electrons, for the group 3 element, provide some shielding
- the overall effect is for the ionisation energy to decrease.

From group 5 to group 6, phosphorus to sulfur

The explanation for the decrease is:

- despite the increased nuclear charge, the electron from the group 6 element is being removed from a p^4 configuration:

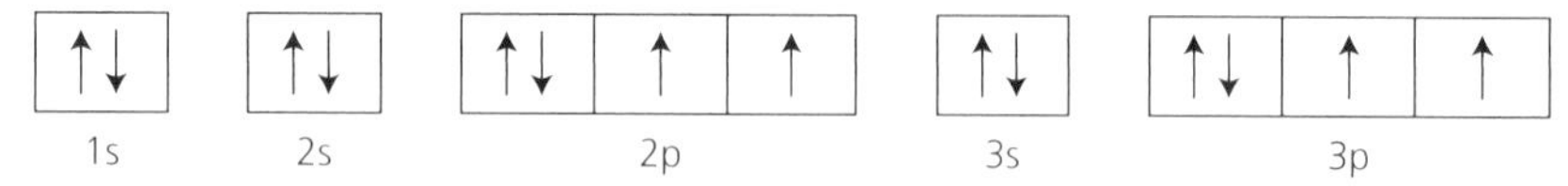

Figure 1.6 Arrangement of electrons in orbitals

- there are four electrons in a p-sub-level, so two of these must be paired in one orbital
- this electron–electron repulsion lowers the attraction between the fourth electron and the nucleus
- so the electron is easier to remove.

Exam practice

1 **(a)** Write the electron configuration of an Al^+ ion. [1]

(b) (i) State the meaning of the term 'first ionisation energy'. [2]

(ii) Write an equation, including state symbols, to show the reaction that occurs when the second ionisation energy of aluminium is measured. [2]

(iii) Explain why the second ionisation energy of aluminium is higher than the first ionisation energy of aluminium. [2]

(c) State and explain the general trend in the first ionisation energies of the period 3 elements sodium to chlorine. [3]

(d) Explain why sulfur has a lower first ionisation energy than phosphorus. [2]

(e) Explain why argon ($Z = 18$) has a much higher first ionisation energy than potassium ($Z = 19$) even though potassium has a larger nuclear charge. [2]

Answers and quick quiz 1 online

Online

Examiners' summary

You should now have an understanding of:

- the properties of protons, neutrons and electrons
- isotopes
- how a mass spectrometer works
- how a mass spectrum can be interpreted to provide information about isotopes
- how to calculate the relative atomic mass of an element using its mass spectrum
- electronic arrangement in terms of s, p, d notation
- ionisation energy
- the evidence for energy levels and sub-levels from ionisation energies

2 Amount of substance

The mole and the Avogadro constant (*L*)

Definitions

Revised

1 mole is defined as the amount of substance that contains as many elementary particles as there are atoms in exactly 12 g of the ^{12}C isotope.

The number 6×10^{23} is referred to as the **Avogadro constant**. It is the number of specified particles — electrons, atoms, molecules or ions — in 1 mole.

The mass of **1 mole** of an element or compound is its relative atomic mass (A_r) or relative molecular mass (M_r) expressed in grams.

An **amount of substance** is measured in moles.

$$\text{number of moles} = \frac{\text{mass of substance}}{\text{mass of 1 mole of that substance}}$$

or mass of substance = mass of 1 mole × number of moles

Calculate the mass of 0.56 moles of potassium dichromate(VI) ($K_2Cr_2O_7$).

Answer

Mass of 1 mole of $K_2Cr_2O_7$ = (2 × 39.0) + (2 × 52.0) + (7 × 16.0) = 294 g

Mass of substance = mass of 1 mole × number of moles

0.56 moles will have a mass of: 294 g × 0.56 = 164.64 g

Examiners' tip

The first step is always to work out the mass of 1 mole. This is then multiplied by the number of moles to calculate a mass.

Empirical and molecular formulae

Definitions

Revised

An **empirical formula** is a formula representing the simplest whole number ratio of atoms of each element in a compound.

A **molecular formula** represents the actual number of atoms of each element in one molecule.

The molecular formulae for ethane and hydrogen peroxide are C_2H_6 and H_2O_2 respectively, whereas the simplest whole number ratios for these formulae — the empirical formula — are CH_3 and HO respectively.

Example 1

A hydrocarbon contains 2.51 g of carbon and 0.488 g of hydrogen. What is the empirical formula of the hydrocarbon?

Answer

mass of carbon = 2.51 g   mass of hydrogen = 0.488 g

Convert into moles by dividing by the A_r for each element:

moles of C $= \frac{2.51}{12.0}$ moles of H $= \frac{0.488}{1.0}$

$= 0.209$ moles $= 0.488$ moles

Simplify the ratio by dividing each by the smaller number of moles:

$\frac{0.209}{0.209} = 1$ $\frac{0.488}{0.209} = 2.33$

Therefore the ratio C:H is 1:2.33 or 3:7 (by multiplying each by 3 to get whole numbers). So the empirical formula is C_3H_7. If the relative molecular mass was given as 86 then the molecular formula would be C_6H_{14} since 2 'lots' of the empirical formula would be required to give this molar mass.

Examiners' tip

Try to recognise certain ratios as being whole number ratios in 'disguise', for example, 1:1.5 is 2:3; 1:1.33 is 3:4 and so on.

Questions may also be set featuring percentage compositions.

Example 2

A compound was found to contain 40.0% sulfur and 60.0% oxygen. What is the empirical formula of the compound? [A_r data: S = 32.1; O = 16.0]

Answer

If we assume the total mass of the compound is 100 g, then the masses of sulfur and oxygen will be 40.0 g and 60.0 g respectively.

mass of sulfur = 40.0 g   mass of oxygen = 60.0 g

Convert into moles: $\frac{40.0}{32.1} = 1.25$ moles $\frac{60.0}{16.0} = 3.75$ moles

The ratio of sulfur to oxygen is 1.25:3.75 or 1:3, so the empirical formula of the compound is SO_3

Examiners' tip

Remember to divide each amount in moles by the smallest value — this will normally give new numbers that are easier to recognise as whole numbers.

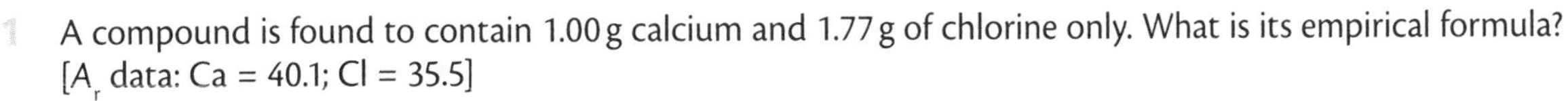

Now test yourself   Tested

1. A compound is found to contain 1.00 g calcium and 1.77 g of chlorine only. What is its empirical formula? [A_r data: Ca = 40.1; Cl = 35.5]
2. A compound of calcium, silicon and oxygen is found to contain 0.210 g of calcium, 0.147 g of silicon and 0.252 g of oxygen. What is its empirical formula? [A_r data: Si = 28.1; Ca = 40.1; O = 16.0]
3. An oxide of nitrogen was found to contain 30.4% nitrogen by mass. The M_r of the oxide is 92.0. [A_r data: N = 14.0; O = 16.0]
   (a) What is the empirical formula of the oxide?
   (b) What is the molecular formula of the compound?

Answers on p. 103

The ideal gas equation

Calculations

Revised

This states that $pV = nRT$, where p is the pressure measured in Pascals (Pa), V is the volume measured in m^3, n is the number of moles; R is the gas constant ($8.31\,J\,K^{-1}\,mol^{-1}$) and T is the temperature in Kelvins.

Examiners' tip

It is essential that the quantities in this equation have the correct units — notice in particular that volume of gas is in m^3, not cm^3 or dm^3.

In an experiment, 0.700 mol of CO_2 was produced. This gas occupied a volume of $0.0450\,m^3$ at a pressure of 100 kPa. Calculate the temperature of the CO_2 and state the units of your answer.

Answer

Using $pV = nRT$

$p = 100\,000\,Pa$, $V = 0.0450\,m^3$, $n = 0.700\,mol$ and $R = 8.31\,J\,K^{-1}\,mol^{-1}$.

Substituting gives:

$100\,000 \times 0.0450 = 0.700 \times 8.31 \times T$

Rearranging gives:

$$T = \frac{(100\,000 \times 0.0450)}{(0.700 \times 8.31)} = 774\,K \text{ (to 3 significant figures)}$$

This temperature, in Celsius, would be $774 - 273 = 501°C$.

1. A 0.905 mol sample of hydrogen gas, H_2, occupies a volume of $0.0330\,m^3$ at a temperature of 200°C. What is the pressure exerted by the gas? The gas constant, R is $8.31\,J\,K^{-1}\,mol^{-1}$.

The mole and reactions

Questions on the mole are asked on many examination papers, so it is essential that this type of question is mastered.

Mass calculations

Revised

Example

23.0 g of calcium carbonate decomposes fully on heating. Calculate the mass of carbon dioxide gas that forms. [A_r data: C = 12.0; Ca = 40.1; O = 16.0]

Answer

The equation for the reaction taking place is $CaCO_3(s) \rightarrow CaO(s) + CO_2(g)$

Step 1

Calculate the number of moles of calcium carbonate.

Molar mass of $CaCO_3 = 40.1 + 12.0 + (3 \times 16.0) = 100.1$

$$\text{Number of moles of } CaCO_3 = \frac{\text{mass used (g)}}{\text{mass of 1 mole (g)}} = \frac{23.0}{100.1} = 0.230\,\text{mol}$$

Step 2

From the equation, $CaCO_3(s) \rightarrow CaO(s) + CO_2(g)$, 1 mole of $CaCO_3$ gives 1 mole of CO_2; the ratio is 1 : 1. So, the amount of CO_2 that forms is also 0.230 moles.

Step 3

Calculate the mass of carbon dioxide.

Mass of CO_2 = mass of 1 mole of CO_2 × number of moles

$= (12.0 + 16.0 + 16.0) \times 0.230 = 10.1$ g of CO_2 gas

Examiners' tip

Remember the three steps:

- **moles** (work out the moles)
- **ratio** (using the balanced symbol equation)
- calculate the **mass** of product (using moles × mass of 1 mole)

Now test yourself

5 Use the equation $Mg(s) + 2HCl(aq) \rightarrow MgCl_2(aq) + H_2(g)$ to determine the volume of hydrogen produced at 25°C and 100 kPa when 2.00 g of magnesium is added to excess hydrochloric acid. [A_r data: Mg = 24.3; H = 1.0]

Answers on p. 103

Tested

Gas calculations

Revised

Equal volumes of gases contain equal numbers of particles, hence the amounts (in moles) will be the same.

Revision activity

Using the internet, find out about the famous French chemist and physicist Joseph Louis Gay-Lussac.

In the following reactions, 100 cm³ of hydrogen is reacted with (a) chlorine and (b) oxygen in reactions that go to completion (no reactants are left). What volumes of gases were formed in each reaction?

Answer

(a) $H_2(g) + Cl_2(g) \rightarrow 2HCl(g)$

From the equation, 1 volume of H_2 reacts with 1 volume of Cl_2 to form 2 volumes of HCl.

So, 100 cm³ of H_2 reacts with 100 cm³ of Cl_2 to form 200 cm³ of HCl

(b) $2H_2(g) + O_2(g) \rightarrow 2H_2O(g)$

2 volumes of hydrogen react with 1 volume of oxygen to form 2 volumes of water vapour.

So, 100 cm³ of hydrogen react with 50 cm³ of oxygen to form 100 cm³ of water vapour.

What volume of hydrogen will react with 150 cm³ of nitrogen in the following reaction, assuming the reaction goes to completion?

$$N_2(g) + 3H_2(g) \rightarrow 2NH_3(g)$$

Other reactions involving gases

Revised

Calculate the volume of gas produced at 298 K and 100 kPa when 1.45 g of lithium metal reacts with water. [A_r data: Li = 6.9]

Answer

$2Li(s) + 2H_2O(l) \rightarrow 2LiOH(aq) + H_2(g)$

Amount of lithium = $\frac{1.45}{6.9}$ = 0.210 moles

Amount of hydrogen produced will be $\frac{0.210}{2}$ (according to the ratios in the equation).

So amount of hydrogen = 0.105 moles

Volume of hydrogen = $\frac{(nRT)}{p} = \frac{(0.105 \times 8.31 \times 298)}{100\,000}$

$= 2.602 \times 10^{-3}\,m^3$ or 2602 cm³

Now test yourself Tested

7 Calculate the volume of oxygen produced at 25°C and 100 kPa when 170 g of hydrogen peroxide decomposes according to the equation:

$$2H_2O_2(aq) \rightarrow 2H_2O(l) + O_2(g)$$

Answers on p. 103

Solutions

Revised

The basic relationship for the amount of a substance in a solution can be expressed as:

$$\text{amount of solute dissolved (moles)} = \frac{\text{volume of solution (cm}^3) \times \text{concentration (mol dm}^{-3})}{1000\,\text{cm}^3}$$

It is also possible to convert everything into dm^3 and use:

$$\text{amount of solute dissolved (moles)} = \text{volume of solution (dm}^3) \times \text{concentration (mol dm}^{-3})$$

Examiners' tip

Sometimes the term 'mol' is used to abbreviate the term 'moles', especially when it follows a number.

Example

Calculate the number of moles of acid dissolved in 25.50 cm^3 of 2.50 × 10^{-3} mol dm^{-3} sulfuric acid.

Answer

$$\text{Number of moles} = \frac{25.50\,\text{cm}^3}{1000\,\text{cm}^3} \times 2.5 \times 10^{-3}\,\text{mol dm}^{-3}$$

$$= 6.375 \times 10^{-5} \text{ moles of } H_2SO_4$$

Now test yourself

8 How many moles of solute are dissolved in:

(a) 10.0 cm^3 of 0.200 mol dm^{-3} NaOH

(b) 250 cm^3 of 1.20 mol dm^{-3} HNO_3?

Answers on p. 103

Solutions and reactions

Many reactions are carried out in solution and calculations involving reacting amounts and volumes of solutions are common in examinations.

Revision activity

On a revision card, write formulae for how the amount of substance can be worked out for solids, solutions and gases. You will find these formulae very useful.

Examiners' tip

The stages involved in this type of calculation are the same as with mass calculations — work out the moles followed by the reaction ratio, and finally the volume (or concentration).

Calculate the volume of 0.200 mol dm^{-3} sulfuric acid required to react exactly with 10.5 cm^3 of 0.400 mol dm^{-3} sodium hydroxide solution.

Answer

$2NaOH(aq) + H_2SO_4(aq) \rightarrow Na_2SO_4(aq) + 2H_2O(l)$

number of moles of sodium hydroxide $= \frac{10.5}{1000} \times 0.400 = 4.2 \times 10^{-3}$ moles

According to the equation, 4.20×10^{-3} mol of sodium hydroxide reacts with ½(4.2×10^{-3}) moles of sulfuric acid (the reacting ratio according to the equation is 2 : 1). So, 2.1×10^{-3} moles of sulfuric acid are required. If the starting concentration is 0.2 mol dm^{-3}, then

2.10×10^{-3} moles $= \frac{\text{volume in cm}^3}{1000\ \text{cm}^3} \times 0.200$ mol dm^{-3}

Rearranging, the volume is calculated as $\frac{2.10 \times 10^{-3} \times 1000}{0.200} = 10.5$ cm^3

Revision activity

Write out the key stages involved in carrying out an acid–base titration. Find out why this method of working is considered to be very accurate.

Calculate the volume of 0.0500 mol dm^{-3} NaOH that will react exactly with 20.0 cm^3 of 0.900 mol dm^{-3} HCl.

Percentage atom economy

Revised

Sustainable development involves maximising our use of the resources available and reducing waste products if at all possible.

In a reaction, a measure of how much of the total mass of reactants is converted into the desired product is called the **atom economy** of that reaction and is defined as:

$$\frac{\text{maximum mass of product}}{\text{total mass of all products}} \times 100$$

Or as

$$\frac{\text{mass of product}}{\text{total } M_r \text{ values for all reactants}} \times 100$$

Examiners' tip

Symbols like A_r and M_r mean *relative* atomic mass and *relative* molecular mass respectively. Neither term has any units.

Calculate the atom economy for ethanol formation in this process:

$$C_6H_{12}O_6(aq) \rightarrow 2C_2H_5OH(aq) + 2CO_2(g)$$

Answer

M_r of glucose = 180, so the percentage atom economy will be

$\frac{2 \times 46}{180} \times 100 = 51.1\%$

What is the atom economy for the formation of chloromethane in the reaction

$$CH_4(g) + Cl_2(g) \rightarrow CH_3Cl(g) + HCl(g)$$

M_r for CH_3Cl is 50.5 and for HCl it is 36.5

Percentage yield

Revised

In a reaction, the amount of product is called its **yield**:

$$\% \text{ yield} = \frac{\text{the mass of product formed}}{\text{the maximum theoretical mass of product}} \times 100$$

Example

When a 15.0 g sample of magnesium is heated in oxygen, it is found that 22.2 g of magnesium oxide forms. What is the percentage yield? [A_r data: Mg = 24.3; O = 16.0]

Answer

Write the chemical equation for the reaction:

$$2Mg(s) + O_2(g) \rightarrow 2MgO(s)$$

Calculate the actual mass of magnesium oxide expected:

amount of magnesium = $\frac{15.0}{24.3} = 0.617$ moles

amount of MgO formed = 0.617 moles

expected mass of MgO = 0.617 × 40.3 = 24.9 g

The % yield is therefore $\frac{22.2}{24.9} \times 100 = 89.2\%$

Chemical formulae and writing chemical equations

Formulae of common ions

Revised

The ability to write correct formulae is an important skill and should be practised. The same is true when writing chemical equations.

Table 2.1 lists some formulae of common ions that are worth knowing.

Table 2.1 Some common ions worth remembering

Formula of ion	Name of ion
CO_3^{2-}	Carbonate
SO_4^{2-}	Sulfate
NO_3^-	Nitrate
NH_4^+	Ammonium
OH^-	Hydroxide
SO_3^{2-}	Sulfite
NO_2^-	Nitrite
HCO_3^-	Hydrogencarbonate
SiO_3^{2-}	Silicate
ClO_3^-	Chlorate
PO_4^{3-}	Phosphate

Revision activity

On a revision card, write the important ions in this table and try to remember them. Keep looking at the card until you know them all. Alternatively, you could write the formulae of the ions into a simple database and use it to test yourself.

You can work out the formulae for some common compounds using a periodic table for reference.

What is the formula for each of these compounds:

(a) magnesium oxide
(b) copper(I) sulfide
(c) manganese(II) nitrate
(d) ammonium sulfate?

Answer

(a) Magnesium oxide is made up of Mg^{2+} and O^{2-} ions. The charges will cancel so the formula is MgO.
(b) Copper(I) sulfide is made of Cu^{+} and S^{2-} ions, swapping over the charges (or valencies) gives Cu_2S.
(c) Manganese(II) nitrate is made of Mn^{2+} and NO_3^- ions. Swapping over the numbers gives $Mn(NO_3)_2$. Remember to use a bracket when multiplying more than one element by a number.
(d) Ammonium sulfate is made of ammonium ions, NH_4^+, and sulfate ions, SO_4^{2-}. Swapping over the valencies gives $(NH_4)_2SO_4$.

Examiners' tip

When deducing the formula for a compound made of ions, simply swap over the charges, or valencies, in their lowest ratio.

What is the formula for each of these compounds?

(a) sodium fluoride
(b) potassium sulfate
(c) aluminium hydroxide?

Table 2.2 lists some other important substances that have formulae that are easier to know, rather than work out.

Table 2.2

Substance	Formulae
Sulfuric acid	H_2SO_4
Nitric acid	HNO_3
Hydrochloric acid	HCl
Hydrogen peroxide	H_2O_2
Ammonia	NH_3
Alkanes	C_nH_{2n+2} (general formula)

Writing balanced symbol equations

Revised

Many reactants and products will be evident from the text of examination questions; others you will be expected to know. For example, the reactions of acids are very important — these are summarised in Figure 2.1.

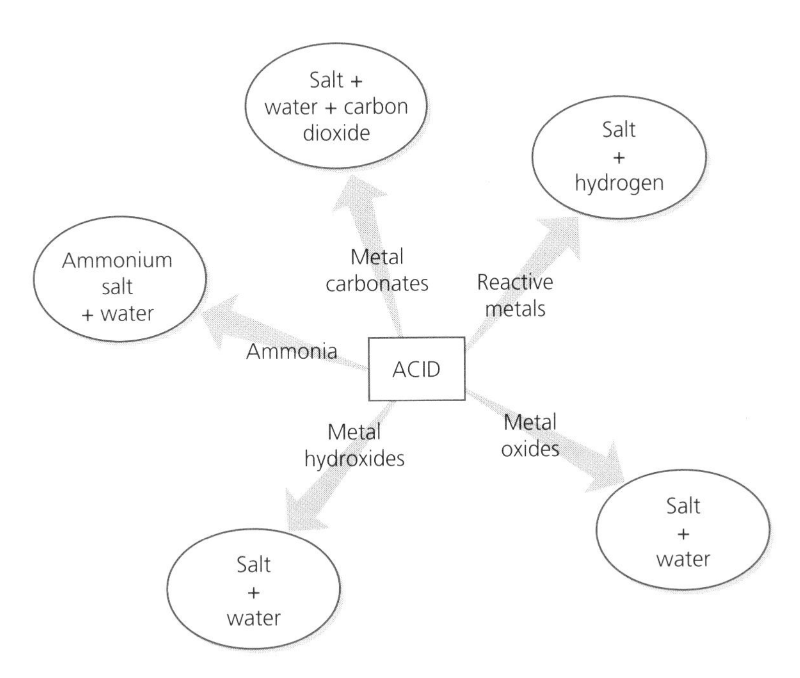

Figure 2.1 Reactions of acids

Example

Write a balanced equation to show how magnesium reacts with dilute nitric acid.

Answer

You need to know that magnesium nitrate and hydrogen gas form. You can then construct an equation and balance it.

- Write the formulae for each of the reactants and products:
  $Mg + HNO_3 \rightarrow Mg(NO_3)_2 + H_2$
- Balance to make it into an equation: $Mg + 2HNO_3 \rightarrow Mg(NO_3)_2 + H_2$
- Add state symbols if they are known:
  $Mg(s) + 2HNO_3(aq) \rightarrow Mg(NO_3)_2(aq) + H_2(g)$

Now test yourself Tested

12 Write a balanced symbol equation for the displacement reaction that takes place between aluminium metal and copper(II) sulfate.

Answers on p. 103

Ionic equations

Revised

Ionic equations are useful for some reactions because they do not include spectator ions — just those ions that do take part in the reaction.

Consider the reaction mentioned previously:

$$Mg(s) + 2HNO_3(aq) \rightarrow Mg(NO_3)_2(aq) + H_2(g)$$

To convert this into an ionic equation, any substance that is dissolved in water has to be written out in its ionic form:

$$Mg(s) + 2H^+(aq) + 2NO_3^-(aq) \rightarrow Mg^{2+}(aq) + 2NO_3^-(aq) + H_2(g)$$

Examiners' tip

It is sometimes easier to remember the common types of ionic equations such as neutralisation $H^+(aq) + OH^-(aq) \rightarrow H_2O(l)$, and the metal/acid reactions as shown.

You then cancel out the spectator ions, in this case the nitrate ions, NO_3^-
This leaves:

$$Mg(s) + 2H^+(aq) \rightarrow Mg^{2+}(aq) + H_2(g)$$

which is the ionic equation for the reaction.

Now test yourself

Tested

13 Write a balanced equation and an ionic equation to show potassium hydroxide solution reacting with hydrochloric acid. Include state symbols in your answer.

Answers on p. 103

Exam practice

1 Fluorine is a pale yellow gas that is known to be extremely reactive. It reacts with magnesium according to the equation:

$$Mg(s) + F_2(g) \rightarrow MgF_2(s)$$

In an experiment, 9.20 g of magnesium react with excess fluorine and magnesium fluoride is formed. [A_r data: Mg = 24.3, F = 19.0]

(a) Calculate the number of moles of magnesium used in the experiment. [1]

(b) Deduce the number of moles of magnesium fluoride formed in the reaction. [1]

(c) Calculate the mass of magnesium fluoride is formed? [1]

(d) What mass of magnesium would be needed if exactly 50.0 g of magnesium fluoride were required? [3]

2 In a titration, 25.00 cm^3 of 0.500 mol dm^{-3} sodium hydroxide is poured into to a conical flask and then titrated with hydrochloric acid of unknown concentration. It is found that 14.50 cm^3 of the acid is required for complete neutralisation. The equation for the reaction is:

$$HCl(aq) + NaOH(aq) \rightarrow NaCl(aq) + H_2O(l)$$

(a) Calculate the number of moles of sodium hydroxide used. [1]

(b) Deduce the number of moles of acid added from the burette. [1]

(c) Calculate the concentration of the dilute hydrochloric acid in mol dm^{-3}. [1]

3 The element indium forms a compound X with hydrogen and oxygen. Compound X contains 69.2% indium and 1.8% hydrogen by mass. Calculate the empirical formula of X. [A_r data: In = 114.8; H = 1.0; O = 16.0] [3]

4 An unknown metal carbonate reacts with hydrochloric acid according to the equation:

$$M_2CO_3(aq) + 2HCl(aq) \rightarrow 2MCl(aq) + CO_2(g) + H_2O(l)$$

A 1.72 g sample of M_2CO_3 was dissolved in distilled water to make 250 cm^3 of solution. A 25.0 cm^3 portion of this solution required 16.6 cm^3 of 0.150 mol dm^{-3} hydrochloric acid for complete reaction.

(a) Calculate the amount of HCl used. Give your answer to 3 significant figures. [1]

(b) Deduce the amount of M_2CO_3 that reacted with the hydrochloric acid. [1]

(c) How many moles of M_2CO_3 must have been present in the 250 cm^3 of solution? [1]

(d) Calculate the relative formula mass, M_r, of M_2CO_3. [1]

(e) Deduce the relative atomic mass of metal M. [1]

Answers and quick quiz 2 online

Online

Examiners' summary

You should now have an understanding of:

- what is meant by the term 'mole'
- the Avogadro constant
- empirical formulae and molecular formulae
- the ideal gas equation
- how to carry out solid, gas and solution calculations
- percentage atom economy and its implications
- percentage yield
- how to write chemical equations and ionic equations.

3 Bonding

Nature of ionic, covalent and metallic bonds

All types of chemical bond — whether ionic, covalent or metallic — are due to electrostatic attractions between positively charged and negatively charged particles.

Ionic bonding

Revised

Ionic bonding happens between oppositely charged **ions** in a lattice.

Dot-and-cross diagrams (Figure 3.1) show how atoms form ions when electrons are **transferred** from a metal atom (which becomes a positively charged ion) to a non-metal atom (which becomes a negatively charged ion).

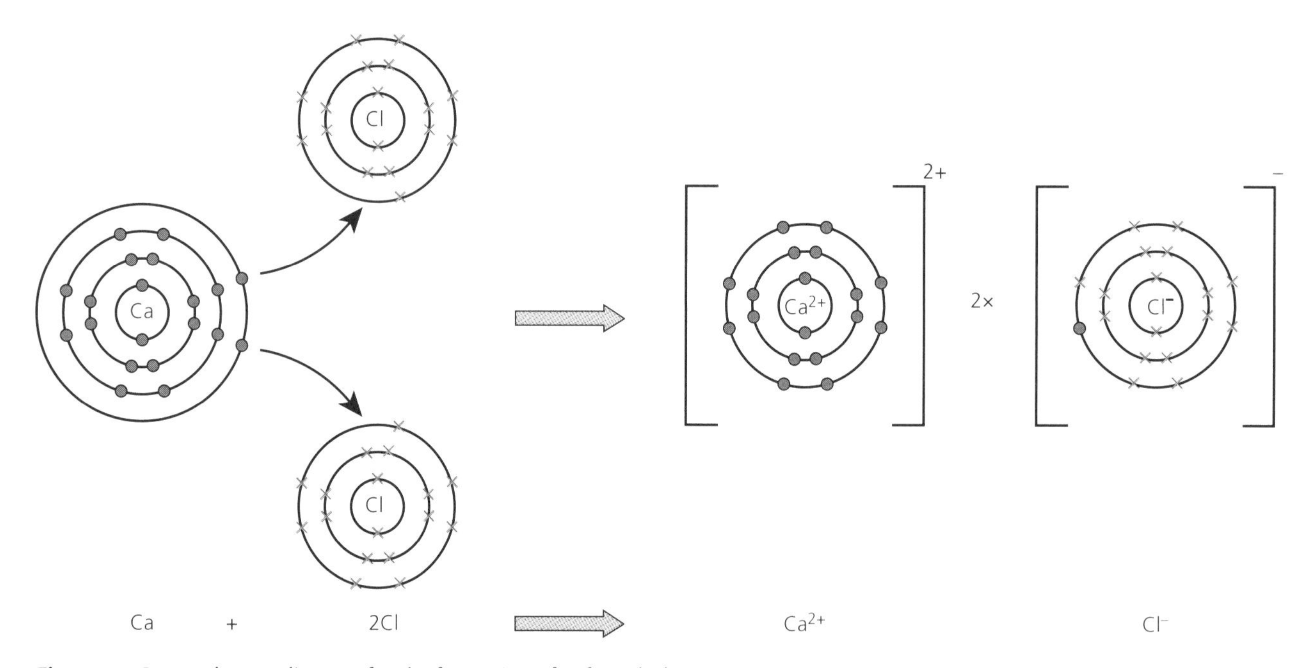

Figure 3.1 Dot-and-cross diagram for the formation of Ca^{2+} and Cl^{-}

Oppositely charged ions are attracted to each other by strong electrostatic forces. This makes it difficult to separate the ions from each other. Therefore ionic substances:

- have high melting points
- are good electrical conductors when molten and in solution (when the ions are free to move) but poor electrical conductors when solid (when the ions are fixed in their lattice and are not free to move).

When ions pack together, they form a giant ionic structure in which positive and negatively charged ions are arranged in a lattice.

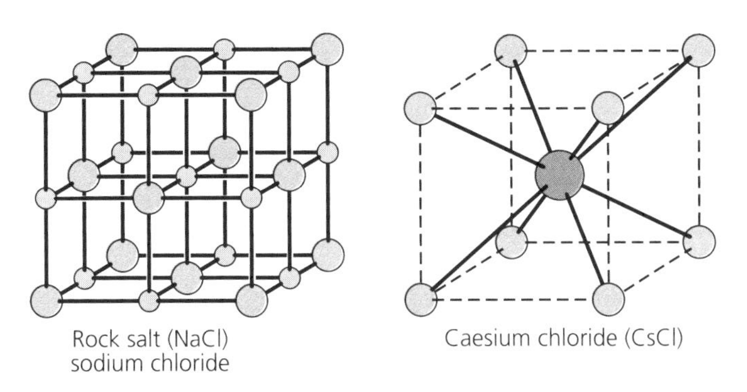

Figure 3.2 Diagrams showing part-structures for NaCl and CsCl

Typical mistake

What is wrong with this student answer? 'Sodium chloride conducts electricity when dissolved in water because electrons are free to move.'

The student should have said 'ions', not 'electrons'. This is a very common error.

Also, never mention *intermolecular* forces when explaining why sodium chloride has a high melting point — mention strong electrostatic forces acting between ions instead, because there are no molecules in sodium chloride.

Now test yourself

Tested

1 Draw dot-and-cross diagrams to show the bonding in **(a)** magnesium oxide and **(b)** sodium oxide. [Atomic number data: Mg = 12; O = 8; Na = 11]

2 Explain why lithium fluoride has a high melting point.

Answers on p. 103

Covalent bonding

Revised

A **covalent bond** involves a **shared** pair of electrons between atoms. A double bond involves two shared pairs of electrons between atoms.

Atoms may bond together covalently to form **molecules**. These molecules may be very easy to separate because they are attracted to each other by weak intermolecular forces — these give the substance very low melting and boiling points. These substances have **simple covalent** structures.

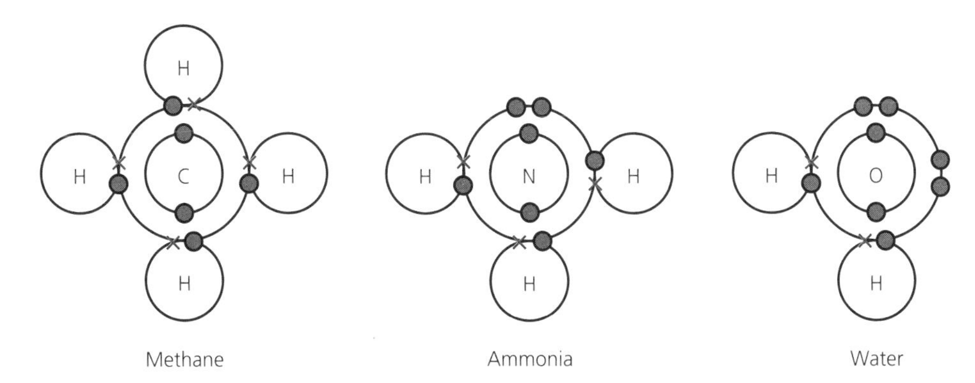

Figure 3.3 Examples of simple covalent structures

Giant molecules, like diamond, may form in which atoms are bonded throughout the structure using strong covalent bonds. This gives substances a very high melting point because these bonds will require a lot of energy to break. These are called **giant covalent (or macromolecular) structures**.

Examiners' tip

Normally it is true that when a metal forms a compound, ionic bonding results; and covalent bonds result when non-metals form a compound. Don't get them round the wrong way!

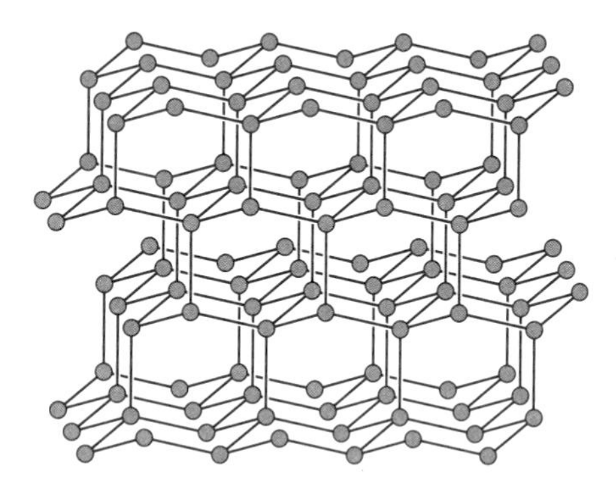

Figure 3.4 Structure of diamond

Draw dot-and-cross diagrams to show the bonding in **(a)** methane and **(b)** nitrogen.
[Atomic number data: C = 6; H = 1; N = 7]

Explain why methane has a low melting point.

Revision activity

Use the internet to look up other small molecules and see if you can draw their electronic structures showing their covalent bonds.

Dative (coordinate) bonding

Revised

A **dative** or **coordinate bond** is one in which both electrons being shared in the covalent bond come from the same atom.

In the formation of the ammonium ion, NH_4^+, the lone pair of electrons on the nitrogen atom is used to form a single covalent bond with a proton:

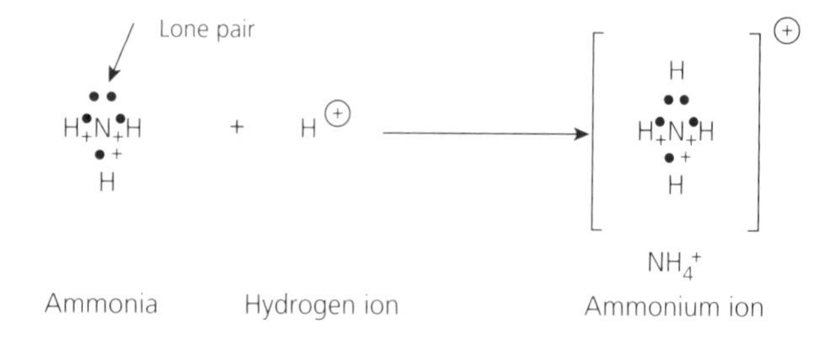

Figure 3.5 Formation of an ammonium ion, NH_4^+

Carbon monoxide and ozone, O_3, are examples of molecules that contain a dative bond in which two electrons are donated from the oxygen atom.

Draw a dot-and-cross diagram for a carbon monoxide molecule.
[Atomic number data: C = 6; O = 8]

Metallic bonding

Revised

Metals consist of a lattice of **positive ions** surrounded by **delocalised electrons** (Figure 3.6)

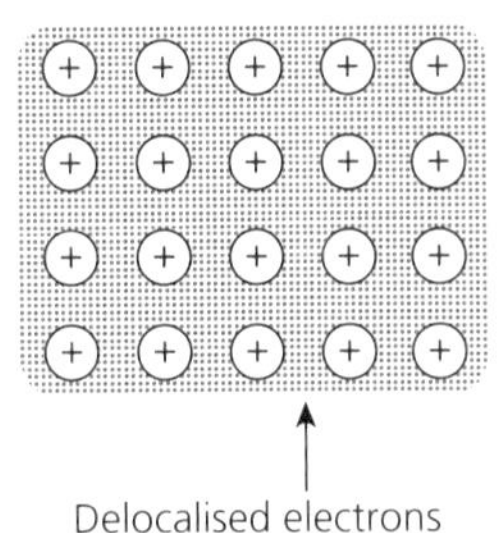

Figure 3.6 Structure of a metal

The mutual attraction of the positive ions for the delocalised electrons is a strong electrostatic force and gives most metals **high melting points** because it is difficult to separate the ions from each other.

The mobile electrons can carry an electrical charge and this explains why metals can conduct electricity so well.

The layers of metal ions within the lattice can also slide across each other fairly easily when a force is applied; this means that metals can be beaten into sheets and are said to **malleable**.

Typical mistake

What is wrong with this student's answer explaining why the electrical conductivity of a metal is good? 'Because there are mobile ions that can carry the charge.'

They should have said that mobile *electrons* are free to carry the charge.

Bond polarity

Electronegativity

Revised

Electronegativity is the power of an atom to withdraw electron density from a covalent bond.

Figure 3.7 shows the graph of electronegativity plotted against atomic number for the first 20 elements.

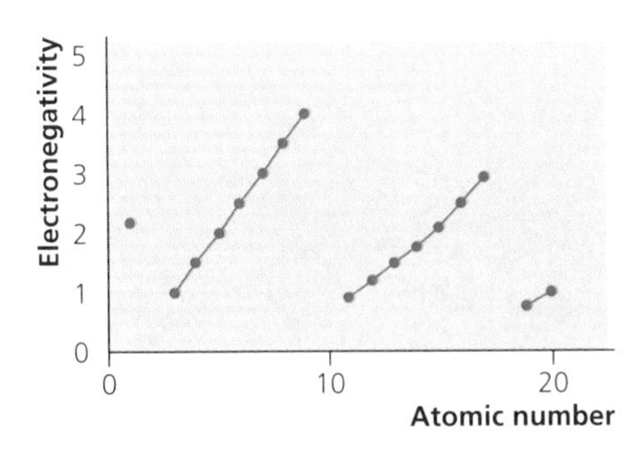

Figure 3.7 How electronegativity varies with atomic number

Electronegativity increases across a period, from left to right, and decreases going down a group. Elements with very high electronegativity values include F, N and O — these elements are very good at attracting bonded electrons towards their nuclei. This attraction causes an **asymmetrical electron distribution** in which the electronegative atom develops a slight negative charge and the other atom develops a slight positive charge. A **polar bond** is produced.

Figure 3.8 shows the electron distribution in a chlorine molecule (a non-polar molecule because both atoms are the same) and hydrogen chloride (a polar molecule due to the different electronegativities of hydrogen (2.1) and chlorine (3.0)). Hydrogen chloride is a polar molecule and has slightly charged ends: $^{\delta+}$H–Cl$^{\delta-}$.

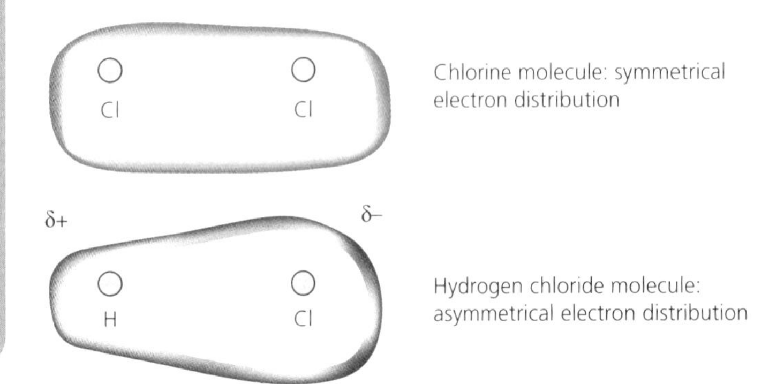

Figure 3.8 Electron distribution in chlorine and hydrogen chloride

Examiners' tip

Elements have different electronegativity values — this means that bonds can differ in their polarity depending on the atoms at the ends of the bond.

Explain why hydrogen fluoride, HF, is a polar molecule whereas hydrogen, H_2, is a non-polar molecule. The electronegativity values of fluorine and hydrogen are 4.0 and 2.1 respectively.

Some molecules may have polar bonds but the molecule still has no overall **dipole moment**. This is because the individual dipoles cancel out because of the three-dimensional shape of the molecule. An example of this is carbon dioxide (Figure 3.9) in which the carbon–oxygen dipoles cancel.

$$\overset{\delta-}{O}=\overset{\delta+}{C}=\overset{\delta-}{O}$$

Figure 3.9 Carbon dioxide molecule

Forces acting between molecules

Forces acting between molecules are called **intermolecular forces**. There are three main types and their strength decreases in the order:

hydrogen bonding > dipole–dipole forces > van der Waals' forces

Molecules with permanent dipoles

Revised

When a molecule has a permanent dipole, either dipole–dipole attractions or hydrogen bonding are possible intermolecular forces.

Examiners' tip

Forces that exist **within** a molecule are strong covalent bonds. Those **between** molecules are weak intermolecular forces.

Hydrogen bonding

If nitrogen, oxygen or fluorine atoms are covalently bonded to hydrogen atoms in a molecule, for example in ammonia or water, an intermolecular attraction called a **hydrogen bond** can act between molecules.

A lone pair of electrons on an oxygen atom of one water molecule is attracted to the slight positive charge on a hydrogen atom in a neighbouring water molecule (Figure 3.10).

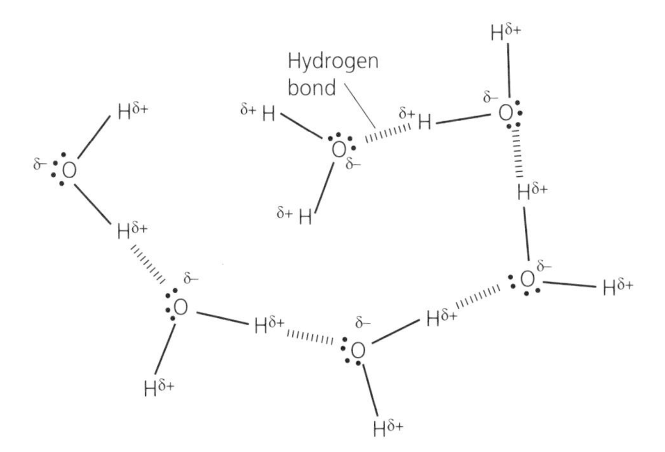

Figure 3.10 Hydrogen bonding in water

Water is a substance that has some anomalous physical properties that can be explained using hydrogen bonding:

- higher melting and boiling points than expected
- lower density as a solid than as a liquid
- the existence of surface tension.

Now test yourself Tested

7 Draw a diagram to show the hydrogen bonding that takes place between two molecules of ammonia. Make sure that you include all lone pairs and partial charges in your diagram.

Answers on p. 104

Dipole–dipole attractions

Polar molecules that do not have N, O or F atoms bonded directly to hydrogen atoms can attract each other using **dipole–dipole attractions**. This is a weaker intermolecular force than hydrogen bonding.

Molecules of hydrogen chloride (Figure 3.11) can attract each other in this way — the slight negative charge on the chlorine atoms attract the slight positive charge of hydrogen atoms on another molecule.

$$\overset{\delta+}{H}\text{—}\overset{\delta-}{Cl} \text{ - - - - - } \overset{\delta+}{H}\text{—}\overset{\delta-}{Cl}$$

Figure 3.11 Why molecules of hydrogen chloride interact

Molecules with temporary dipoles

Revised

Many molecules are non-polar and yet they are still able to attract each other using **van der Waals' forces**, the weakest of all intermolecular forces.

The electrons in all molecules are in constant motion and this movement causes 'wobbles' in electron clouds that result in temporary dipoles. This dipole may then induce another temporary dipole in a neighbouring molecule. The attraction between these temporary dipoles is called the van der Waals' force.

The size of van der Waals' forces depends on the number of electrons in a molecule and also the area of contact of one molecule with another.

Explain why the boiling points of the hydrocarbons — methane, ethane, propane, butane and pentane — increase in the order written.

Answer

As the relative molecular mass increases, the number of electrons in each molecule increases. This means that more and stronger van der Waals' forces exist between the molecules, and they will be more difficult to separate from each other, giving them higher boiling points.

Examiners' tip

The melting and boiling points of a homologous series of hydrocarbons and the elements in group 7 (the halogens) and group 0 (the noble gases) all increase as relative molecular mass increases. This is due to the increasing number of electrons and therefore stronger van der Waal's forces.

8 Explain why the boiling points of the halogens increase on descending the group.

Answers on p. 104

States of matter

Structure of substances

Revised

Substances have different physical properties that depend on their structures (Figure 3.12).

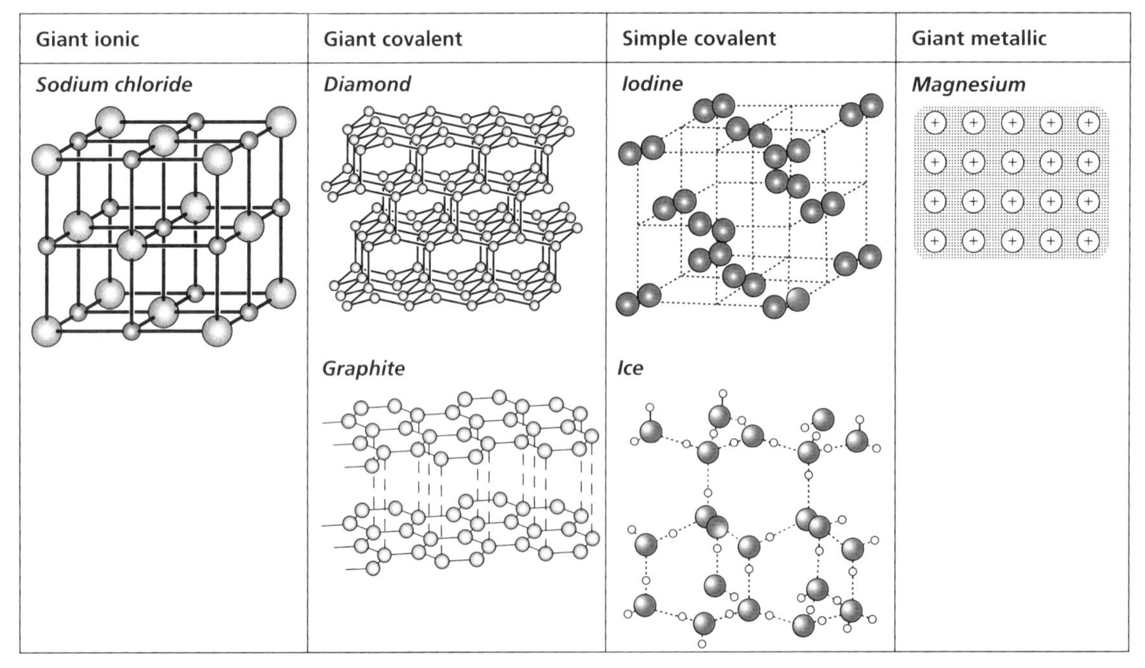

Figure 3.12 Examples of substances with different structures

Giant ionic substances have high melting and boiling points because the oppositely charged ions are attracted to each other by strong electrostatic forces. They are good electrical conductors when molten and in solution because the ions are free to carry the electrical charge. In the solid state, the ions are fixed in place in the lattice.

Giant covalent (macromolecular) substances have high melting and boiling points because the atoms are bonded throughout by strong covalent bonds. This means that a lot of energy is required to break the bonds and to melt the substance.

Simple covalent substances have low melting and boiling points because it is relatively easy to overcome the weak intermolecular forces to separate one molecule from another.

Metallic substances generally have high melting and boiling points because there are strong electrostatic forces acting between the mobile delocalised electrons and the positive ions in the lattice. They are good electrical conductors because the electrons are free to move and carry the charge.

Now test yourself

Tested ☐

9 Explain why sodium fluoride, NaF, has a high melting point whereas the element fluorine, F_2, has a low melting point.

Answers on p. 104

Shapes of molecules and ions

Deducing the shape of a molecule

Revised ☐

The three-dimensional shape of a molecule depends on the repulsion of the electron pairs around the central atom in the molecule. The decreasing order of strengths of these is

lone pair–lone pair > lone pair–bonding pair > bonding pair–bonding pair

See Table 3.1 on page 34 to help you work out the shapes of molecules.

Table 3.1 The shapes of molecules

Number of bonding pairs around the central atom (or atom–atom links)	Number of non-bonding pairs (lone pairs)	Name of shape with diagram and examples
2	0	Linear 180° CO_2, HCN, $BeCl_2$, $[Ag(NH_3)_2]^+$
2	2	Bent H_2O, H_2S, NO_2^-, ClO_2^-
3	0	Trigonal planar 120° BF_3, NO_3^-, CO_3^{2-}, SO_3
3	1	Pyramidal 107° NH_3, PH_3, SO_3^{2-}, ClO_3^-
3	2	T-shape ClF_3
4	0	Tetrahedral 109°28′ NH_4^+, CH_4, SO_4^{2-}, $CuCl_4^{2-}$
4	2	Square planar XeF_4, ICl_4^-
5	0	Trigonal bipyramidal 120° PCl_5
6	0	Octahedral SF_6, $[M(H_2O)_6]^{2+}$ where M = a metal ion

When working out the shape of a molecule or ion, you must work out the number of bonding electrons pairs and lone pairs.

Example

Deduce the shape of a hydrogen sulfide, H_2S, molecule.

Answer

- Sulfur is in group 6 of the periodic table and so its atoms have 6 outer electrons.
- Each hydrogen atom shares 1 electron to make a single covalent bond.
- This means there are 8 outer electrons in 4 pairs— 2 bonding pairs and 2 lone pairs.
- Lone pairs repel the bonding pairs slightly more strongly than the bonding pairs repel each other, so the internal angle is slightly less than the tetrahedral angle (109.5°).
- The shape name is bent (non-linear) and the internal angle is about 104.5° (see Figure 3.13).

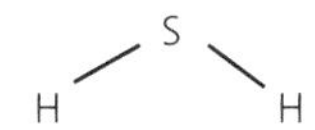

Figure 3.13 A hydrogen sulfide molecule

Examiners' tip

As a 'rule of thumb', each lone pair causes the expected tetrahedral angle to decrease by about 2.5°. So, one lone pair of electrons would cause a decrease of about 5° from 109.5° — hence 104.5° (as in water and hydrogen sulfide).

Example

Deduce the shape of a PF_6^- ion.

Answer

- Phosphorus is in group 5 of the periodic table and so its atoms have 5 outer electrons.
- Each fluorine atom shares 1 electron to make a single covalent bond, contributing 6 electrons to the outer layer.
- The negative charge on the ion adds 1 more electron.
- The total number of electrons around the phosphorus atom is 5 + 6 + 1 = 12, or 6 bonding pairs. There are no lone pairs.
- The shape (Figure 3.14) will therefore be octahedral and the internal bond angles will be 90° and 180°.

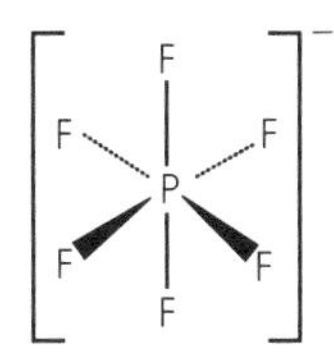

Figure 3.14 A PF_6^- ion

Revision activity

Draw the common shapes in Table 3.1 and use wedges to indicate that the molecules are three-dimensional.

Now test yourself

10 Sketch the shapes and indicate the internal bond angles and shape names of each of these covalent particles:

(a) NH_4^+

(b) BF_3

(c) SF_6

Answers on p. 104

Exam practice

1 **(a)** Describe, with a diagram, the lattice structure of sodium chloride. [2]

(b) Using your diagram from part (a), explain why sodium chloride has the following properties:

(i) Sodium chloride is a good electrical conductor when molten but not when solid.

(ii) Sodium chloride has a high melting point. [2]

2 Explain the meaning of the following terms:

(a) covalent bond [2]

(b) dative bond [2]

(c) giant covalent (macromolecular) structure [2]

3 **(a)** Draw a dot-and-cross diagram for boron trifluoride (BF_3). [2]

(b) Comment on the outer electronic structure of the boron atom in boron trifluoride. [1]

(c) Considering your previous answers, indicate how boron trifluoride may react with ammonia. [2]

4 **(a)** State the strongest type of intermolecular force present in water and the strongest type of intermolecular force in hydrogen sulfide (H_2S). [2]

(b) Draw a diagram to show how molecules of water are attracted to each other by the type of intermolecular force you stated in part (a). In your diagram draw a minimum of three water molecules. [3]

(c) Explain why the boiling point of water is much higher than the boiling point of hydrogen sulfide. [2]

(d) Explain why the boiling points increase from H_2S to H_2Te in group 6. [2]

Answers and quick quiz 2 online

Online

Examiners' summary

You should now have an understanding of:

- ionic bonding
- covalent bonding
- dative (coordinate) bonding
- metallic bonding
- bond polarity and electronegativity
- hydrogen bonding
- dipole–dipole forces
- van der Waals' forces
- the relationship between physical properties and structure
- shapes of molecules and ions

4 Periodicity

Periodicity is the regular and **repeating pattern** of various physical or chemical properties. Since the outer electronic configuration of atoms is a periodic function, we expect other properties to change accordingly.

Classification of elements in s, p and d blocks

The periodic table

Revised

The periodic table is an arrangement of the known chemical elements according to their atomic numbers.

Elements are put into vertical columns called **groups**, in which all elements have similar chemical properties, as well as horizontal rows called **periods**.

Elements are classified as s, p, d or f block elements according to their position in the periodic table. For example, group 1 and 2 elements are in the s block (see Figure 4.1) and their outer, most high energy electrons, occupy the s-sub-level.

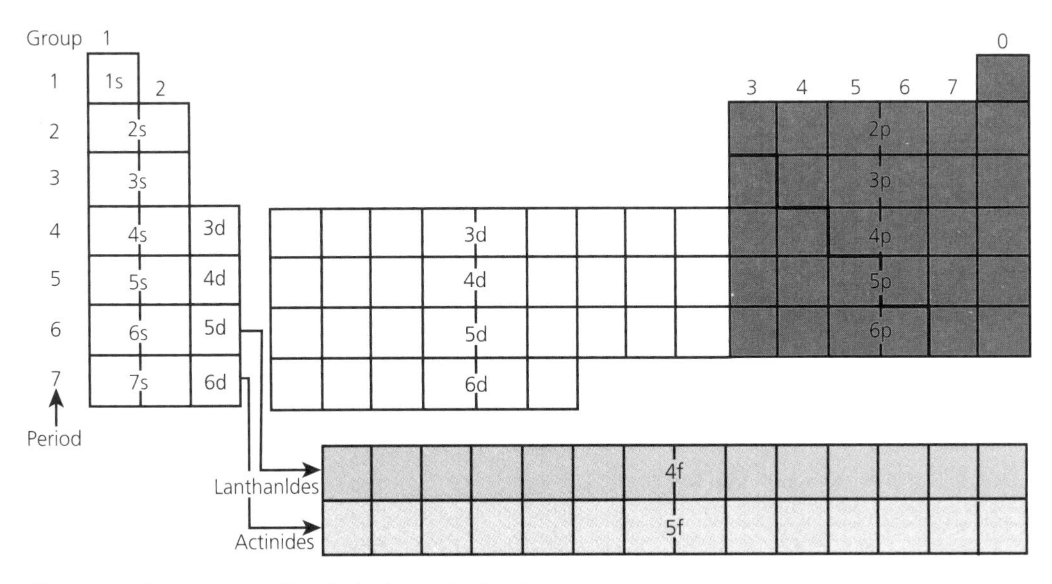

Figure 4.1 The s, p, d and f blocks in the periodic table

Now test yourself

1. In which block of the periodic table are the elements in groups 7 and 0 to be found?
2. The electronic configuration of boron, B, is $1s^2, 2s^2, 2p^1$. In which block of the periodic table would you expect boron be to be found?

Answers on p. 104

Properties of the period 3 elements

Atomic radius

Revised

Figure 4.2 shows atomic radius plotted against atomic number for periods 2–4 in the periodic table. A regular repeating pattern is seen; this is an example of periodicity.

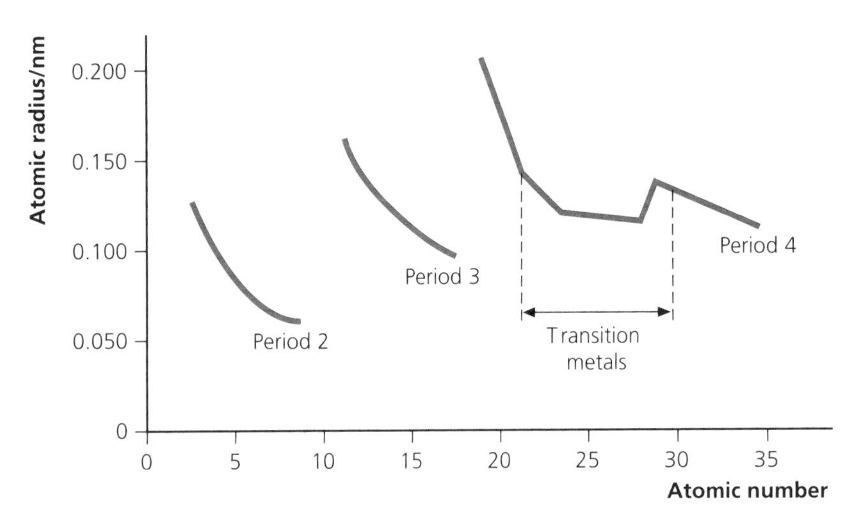

Figure 4.2 Graph of atomic radius versus atomic number

The graph shows two main trends.

Across a period

Moving from left to right across a period, the atomic radius generally decreases.

As the atomic number increases, the number of protons in the nucleus increases. Electrons are also being added, and to the same energy level. There is a greater electrostatic attraction between the outer electrons and the nucleus and so the atomic radius decreases.

Down a group

Electrons are being added to a higher energy level, progressively further from the nucleus. There is a reduction in the electrostatic attraction because of the increasing distance and increased shielding of nuclear attraction by inner levels and so the atomic radius increases.

First ionisation energy

Revised

Figure 1.4 shows how the first ionisation energy changes with atomic number. Chapter 1 outlines the evidence for electron energy levels and sub-levels, and gives an explanation of the ionisation energy trends observed down a group and across a period.

Melting and boiling points

Revised

On moving across the periodic table from left to right along period 3, there is a gradual change in structure of the solid element (Table 4.1). When explaining the trends in melting and boiling points, it is important to know how the structures change.

Revision activity

Using the internet, find the boiling points of the period 2 elements (lithium to neon) and then plot them on a graph to see that the same pattern is seen as with the period 3 elements.

Table 4.1 Element structures in period 3

Group	1	2	3	4	5	6	7	0
Element	Na	Mg	Al	Si	P	S	Cl	Ar
Structural type	Giant metallic			Giant covalent	Simple covalent			Individual atoms

Figure 4.3 shows how the general trend seen in the melting points of the elements across a period relates closely to the trend seen in structure.

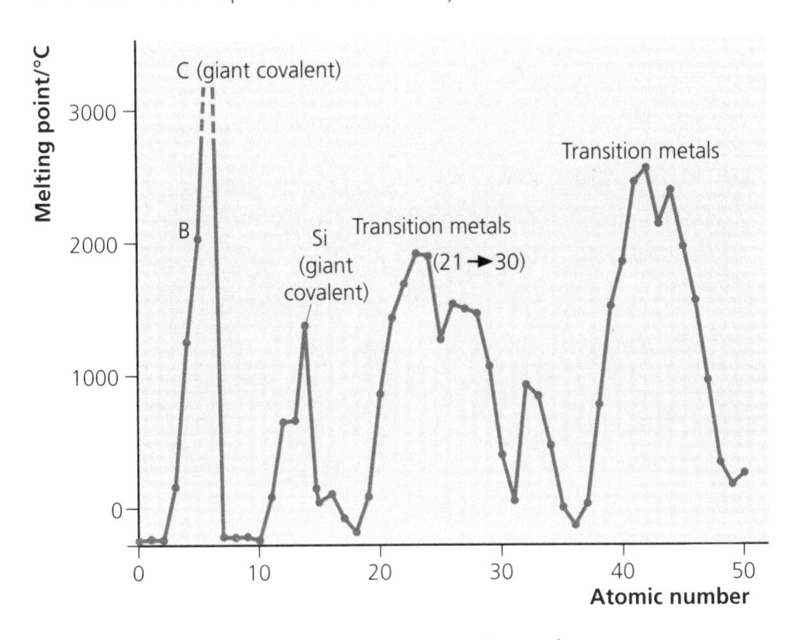

Figure 4.3 Graph of melting point versus atomic number

- The melting and boiling points of metals increases from group 1 through to group 3. This is because there are more electrons being delocalised into the **giant metallic structure**, and so there will be a greater electrostatic attraction between the positive ions and the mobile electrons. For example, the attraction between Al^{3+} ions and three electrons (on average) will be greater than that between Na^{+} ions and one electron. It will therefore be more difficult to separate the ions from each other in the melting or boiling process.
- The group 4 element has the highest melting and boiling point in period 3. Silicon atoms are strongly covalently bonded throughout in a **giant covalent structure** similar to that of diamond, so it will be difficult to melt or boil the element because this will involve breaking strong covalent bonds.

- There is a large drop after silicon but from then to the end of the period the melting point depends only on van der Waals' forces, which in turn depend on the size of the molecules involved. In group 5, phosphorus consists of separate P_4 molecules in a **simple covalent structure**. These molecules will be relatively easy to separate because their weak van der Waals' forces will be easy to overcome. Sulfur in group 6 consists of S_8 molecules, chlorine in group 7 consists of Cl_2 molecules, and argon in group 0 consists of separate atoms. Larger molecules with more electrons, for example S_8, give rise to larger van der Waals' forces and hence their melting and boiling points are higher because it requires more energy to separate the molecules from each other.

Tested

Explain the following:

(a) Why does sodium have a larger atomic radius than magnesium?

(b) Why does aluminium have a lower first ionisation energy than magnesium?

(c) Why does sulfur have a higher boiling point than chlorine?

Answers on p. 104

Exam practice

1 An element X has the following first six ionisation energies in kJ mol^{-1}:

577, 1820, 2740, 11 600, 14 800, 18 400

(a) Explain how you know that element X is in group 3 of the periodic table. [1]

(b) Element Y is in the same group as element X, but it is placed in the period below X in the periodic table.

(i) Give an approximate value for the first ionisation energy of element Y. [1]

(ii) Explain, using ideas of electronic structure, why you expect element Y to have this ionisation energy. [2]

(c) Two elements W and Z are in the same period as X, but W is in the group before X, and Z is in the group after X in the periodic table.

(i) Give approximate first ionisation energies for elements W and Z. [2]

(ii) Explain, using ideas of electronic structure, why elements W and Z have these ionisation energies. [2]

2 Indicate the general trend in the following properties across a period from group 1 to group 7:

(a) **(i)** first ionisation energy [1]

(ii) atomic radius [1]

(b) Explain the variation in melting points across a period. [2]

3 Explain these observations:

(a) Sodium has a larger atomic radius than lithium. [2]

(b) Silicon has a higher melting temperature than phosphorus (P_4). [2]

Answers and quick quiz 3 online

Online

Examiners' summary

You should now have an understanding of:

- what is meant by the term 'periodicity'
- how elements can be classified in s, p, and d blocks in the periodic table
- how atomic radius, first ionisation energy and melting and boiling point can be used to demonstrate periodicity.

5 Alkanes

Fractional distillation of crude oil

Alkanes are **hydrocarbons** — this means that their molecules contain **only** hydrogen and carbon atoms. The general formula of an alkane is C_nH_{2n+2}

Alkanes are **saturated** — this means that they contain only carbon–carbon single bonds.

The alkanes are found in crude oil and are separated from this by a process called **fractional distillation**.

Fractional distillation

Revised

Crude oil is a complex mixture of mainly hydrocarbons. The complex mixture can be separated, by distillation, into smaller mixtures or fractions that have different boiling points. These different components (fractions) of the mixture can be drawn off at different levels in a fractionating column because of the decreasing temperature gradient up the column.

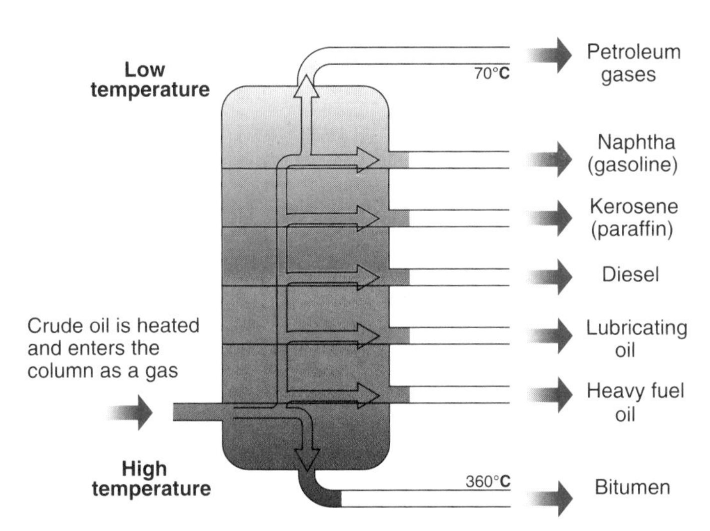

Figure 5.1 Fractional distillation of crude oil

Formulae and naming

Formulae

Revised

The first five members of the alkane homologous series are shown in Figure 5.2.

```
   H
   |
H—C—H
   |
   H
methane

   H  H
   |  |
H—C—C—H
   |  |
   H  H
ethane

   H  H  H
   |  |  |
H—C—C—C—H
   |  |  |
   H  H  H
propane

   H  H  H  H
   |  |  |  |
H—C—C—C—C—H
   |  |  |  |
   H  H  H  H
butane

   H  H  H  H  H
   |  |  |  |  |
H—C—C—C—C—C—H
   |  |  |  |  |
   H  H  H  H  H
pentane
```

Figure 5.2 The first five alkanes

In general, members of a **homologous series**:

- have the same general formula
- contain the same functional groups and have similar chemical properties
- each member differs from the next by a CH_2 unit.

The **molecular formula** gives the actual number of atoms of each element in a molecule. The **empirical formula** gives the simplest whole number ratio of atoms of each element in a molecule.

Using butane as an example:

- the **displayed formula** showing all the bonds is

```
   H  H  H  H
   |  |  |  |
H—C—C—C—C—H
   |  |  |  |
   H  H  H  H
```

- the **structural formula** is $CH_3CH_2CH_2CH_3$
- the **molecular formula** is C_4H_{10} — this shows the actual number of atoms of each element in the molecule, but shows no structure
- the **empirical formula** is C_2H_5 — this shows the simplest whole number ratio for the numbers 4 and 10 in the molecular formula.

Examiners' tip

Ensure that you understand the differences between each kind of formula — they can sometimes look very different.

What is the displayed formula, structural formula, molecular formula and empirical formula for hexane, C_6H_{14}?

Naming alkanes

Revised

Unbranched alkanes are easy to name because they follow the 1: meth-, 2: eth-, 3: prop-, 4: but- system with -ane at the end of the name. For example, the alkane containing a longest chain of three carbon atoms is called propane.

Branched-chain alkanes are named according to the following steps.

1 Find the longest continuous chain of carbon atoms.
2 Look for any branches along the longest chain — how many carbons atoms are there in each branch and at what positions are they along the longest chain?

Example

Name these molecules.

(a)

```
     H  H    H  H
     |  |    |  |
H—C—C —C—C—H
     |  |    |  |
     H  CH3  H  H
```

(b)

```
     H  CH3 H  H  H  H
     |  |   |  |  |  |
H—C—C—C—C—C—C—H
     |  |   |  |  |  |
     H  CH3 H  H  H  H
```

Answer

(a) It has 4 carbon atoms in its longest chain.
It has 1 methyl group positioned at the number 2 carbon atom (counting from the shortest end).
Name: 2-methylbutane

(b) It has 6 carbon atoms in its longest chain.
It has two methyl groups and they are both at the number 2 position.
Name: 2,2-dimethylhexane.

Examiners' tip

The arrangement around each carbon atom is actually tetrahedral but when the structures are 'squashed' onto a flat page, the angles look like 90° or 180° but they are all 109.5°.

Now test yourself

2 Name these alkanes, all of which are isomers of C_5H_{12}

(a)

CH_3—CH_2—CH_2—CH_2—CH_3

(b)

```
     CH3
     |
CH3—CH—CH2—CH3
```

(c)

```
      CH3
      |
CH3—C—CH3
      |
      CH3
```

Answers on p. 104

Tested

Isomerism

Structural isomers

Revised

Many organic molecules, like alkanes, can form more than one structure with the same number of atoms in each molecule — these are called **isomers**.

There are several different types of structural isomerism — **chain, position** and **functional group** isomers.

Structural isomers are molecules with the same molecular formula but different structural formulae.

Chain isomerism

Revised

These isomers arise because of the possibility of branching in carbon chains. For example, there are two isomers of C_4H_{10}. In one of them, the carbon atoms lie in a 'straight' chain whereas in the other the chain is branched.

```
     H  H  H  H              H  H  H
     |  |  |  |              |  |  |
H—C—C—C—C—H          H—C—C—C—H
     |  |  |  |              |  |  |
     H  H  H  H              H  CH3 H
```

butane   2-methylpropane

Position isomerism

Revised

In position isomerism, the basic carbon skeleton remains unchanged, but important groups are attached to different carbon atoms along the chain.

$CH_3—CH_2—CH_2—Br$ (1-bromopropane)

$CH_3—CH(Br)—CH_3$ (2-bromopropane)

Functional group isomerism

Revised

In this type of structural isomerism, the isomers contain different functional groups — that is, they belong to different families of compounds (different homologous series).

For example, a molecular formula C_3H_6O could represent either propanal (an aldehyde) or propanone (a ketone):

$CH_3—CH_2—CHO$ (propanal)

$(CH_3)_2C=O$ (propanone)

Cracking of alkanes

Breaking carbon–carbon bonds

Revised

Longer-chain alkanes are not as marketable as the smaller and more useful alkanes. Because of this low demand, the larger molecules are thermally, or catalytically, cracked to form both higher value alkanes and alkenes.

Cracking involves breaking of carbon–carbon bonds in an alkane, and so energy is required to make this happen. It is the process by which long-chain alkanes are converted into smaller alkanes and, normally, alkenes.

For example, the alkane decane, $C_{10}H_{22}$, could be cracked as follows:

$$C_{10}H_{22}(g) \rightarrow C_8H_{18}(g) + C_2H_4(g)$$

Now test yourself

Write an equation to show the cracking of the alkane nonane, C_9H_{20}, to form ethene and one other product.

Catalytic cracking involves using a slightly increased pressure and a high temperature — a **zeolite catalyst** is often used. Aromatic hydrocarbons and motor fuel are two of the important products formed in this catalytic cracking process.

Thermal cracking involves using a high temperature a high pressure and this process produces a high percentage of **alkenes**.

Examiners' tip

Cracking often produces ethene, C_2H_4 — this can be used to make the polymer poly(ethene).

Combustion of alkanes

Fuels

Revised

Alkanes are often used as fuels when burned in oxygen. The products formed depend on the amount of oxygen gas available. For example, propane gas, C_3H_8, can be combusted in excess oxygen — complete combustion:

$$C_3H_8(g) + 5O_2(g) \rightarrow 3CO_2(g) + 4H_2O(l)$$

Alternatively it can burn by partial or incomplete combustion, in which the oxygen supply is limited:

$$C_3H_8(g) + \frac{7}{2}O_2(g) \rightarrow 3CO(g) + 4H_2O(l)$$

Examiners' tip

Note that water is always formed in hydrocarbon combustion. However, carbon monoxide or carbon can be formed when the oxygen supply is limited.

Now test yourself

Tested

4 If $100\,cm^3$ of ethane is combusted completely, what volume of oxygen is required to ensure that this happens? Assume that all volumes are measure under the same conditions.

5 Write an equation to show the incomplete combustion of methane in which only gases are formed.

Answers on p. 104

Environmental issues

Revised

- In a typical car engine in which diesel or petrol is the fuel, carbon monoxide, unburned hydrocarbons and oxides of nitrogen (NO_x) are found in exhaust gases — all of these are potential pollutants and can present a problem to human health.
- Pollutants such as nitrogen monoxide, NO(g), and carbon monoxide, CO(g), are removed from car exhaust fumes using a catalytic converter. The gases react **together** on the surface of a catalyst such as rhodium, palladium or platinum to form products that are less harmful:

  $$2NO(g) + 2CO(g) \rightarrow 2CO_2(g) + N_2(g)$$
- If fuels contain sulfur, then sulfur dioxide will form when the fuel is burned. This gas is acidic and contributes to acid rain. It can be removed by passing it through powdered calcium oxide (a base) so that it reacts. An acid–base neutralisation occurs:

  $$CaO(s) + SO_2(g) \rightarrow CaSO_3(s)$$
- Gases such as carbon dioxide, water vapour and unburned hydrocarbons such as methane are all **greenhouse gases**. This means that they are able to absorb infrared radiation and this results in the temperature of the atmosphere increasing. This is called **global warming**.

Now test yourself

6 Give the names of two pollutants that are formed when a fuel such as petrol is combusted. How are these pollutants removed?

Answers on p. 104

Tested

Exam practice

1 The alkane pentane is a saturated hydrocarbon.

(a) (i) State the meaning of the terms 'saturated' and 'hydrocarbon' as applied to alkanes. [2]

(ii) Give the general formula for the alkanes. [1]

(b) Pentane burns completely in oxygen.

(i) Write an equation for this reaction. [1]

(ii) State how the products of this reaction may affect the environment. [1]

(c) Give the name of the solid pollutant that may form when pentane burns incompletely in air. [1]

(d) A molecule of C_9H_{20} can be cracked to form pentane and one other product.

(i) Write an equation for this cracking reaction. [1]

(ii) Suggest a type of compound that can be manufactured from the other product of this cracking reaction. [1]

(iii) State why a high temperature is required for cracking reactions to occur. [1]

Answers and quick quiz 3 online

Online

Examiners' summary

You should now have an understanding of:

- what alkanes are
- the terms 'empirical formula', 'molecular formula', 'structural formula' and 'displayed formula'
- homologous series and functional groups
- how to name simple alkanes containing up to six carbon atoms
- what is meant by 'isomers' and, in particular, the various types of structural isomers
- the fractional distillation of crude oil
- the cracking of long-chain alkanes and uses for the products
- the different types of combustion of alkanes
- the consequences of combustion in terms of pollution and how some pollutants are removed

6 Energetics

Enthalpy change, ΔH

An enthalpy change, ΔH, is defined as the heat energy change at constant pressure. If the enthalpy change is a standard value, with the symbol $\Delta H^{\ominus}_{298,}$ then particular conditions are specified: 100 kPa pressure and a temperature of 298 K.

The sign of ΔH

Revised

In a chemical process, heat energy can move either from the surroundings to the chemical system or vice versa. If heat energy moves from the system to the surroundings, the chemicals lose heat energy, an **exothermic reaction** results and the sign of ΔH is negative. If heat energy is transferred from the surroundings to the system, an **endothermic reaction** results and the sign of ΔH is positive.

Calorimetry

Calculations

Revised

Calorimetry is an experimental method for determining enthalpy changes. Simple reactions can be studied in this way.

When calculating the enthalpy change for a reaction, it is usually necessary to convert from a temperature change measurement recorded in an experiment (in which water is heated or cooled) into a heat energy measurement, in joules. This is done using the relationship:

$q = mc\Delta T$

- q = quantity of heat energy, in J
- m is the mass of the water, in g
- c is the specific heat capacity of water, $4.18\,J\,K^{-1}\,g^{-1}$
- ΔT is the temperature change in °C or K

Typical mistake

Many candidates forget that m is the mass of water and not the mass of any solid used in the experiment.

Example 1

Calculate the enthalpy change (in $kJ\,mol^{-1}$) for the following reaction of magnesium (A_r = 24.3).

2.00 g of magnesium powder is added to an excess of 25.0 cm^3 of dilute sulfuric acid solution and the temperature is found to increase by 52.4°C.

Answer

Using $q = mc\Delta T$

$m = 25.0\,g$ $c = 4.18\,J\,K^{-1}\,g^{-1}$ $\Delta T = 52.4°C$

- $q = 25.0 \times 4.18 \times 52.4 = 5476\,J = 5.476\,kJ$
- Amount of magnesium used $= \dfrac{2.00}{24.3} = 0.0823$ moles
- Heat energy released $= \dfrac{q \text{ (in kJ)}}{\text{amount (in moles)}} = \dfrac{5.476}{0.0823} = 66.5\,kJ\,mol^{-1}$
- The reaction is exothermic, so ΔH is negative, so the enthalpy change is $-66.5\,kJ\,mol^{-1}$

Example 2

The following experiment was carried out in which a sample of ethanol ($M_r = 46.0$) was put into a spirit burner (Figure 6.1). The ethanol was ignited and the temperature of the 100 cm^3 of water in the metal calorimeter was found to increase from 20.1°C to 52.6°C. The mass of the spirit burner decreased from 35.64 g to 35.14 g. Calculate the enthalpy of combustion of ethanol.

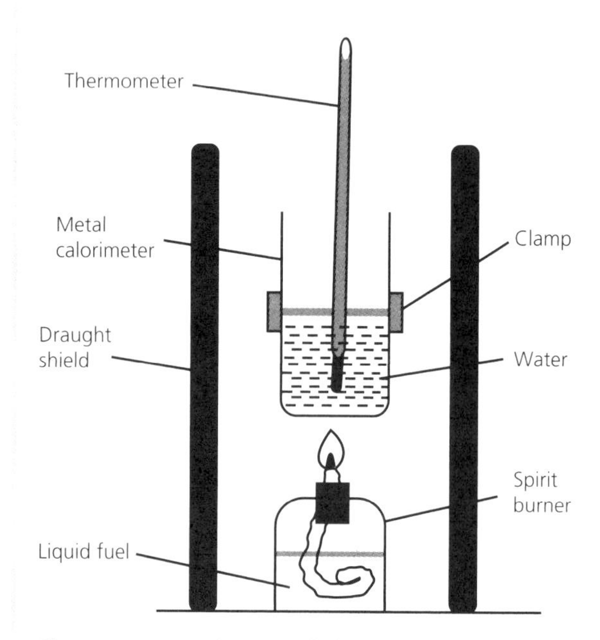

Figure 6.1 Measuring an enthalpy of combustion

Answer

$q = mc\Delta T$

$m = 100.0\,g$ $c = 4.18\,J\,K^{-1}\,g^{-1}$ $\Delta T = 52.6 - 20.1 = 32.5°C$

- $q = 100 \times 4.18 \times 32.5 = 13\,585\,J = 13.59\,kJ$
- Amount of ethanol used $= \dfrac{\text{mass of ethanol combusted}}{M_r \text{ for ethanol}}$

  $= \dfrac{(35.64 - 35.14)}{46.0} = 0.0109$ moles

- Heat energy released $= \dfrac{q \text{ (in kJ)}}{\text{amount (in moles)}} = \dfrac{13.59}{0.0109}$

  $= 1247\,kJ\,mol^{-1}$

- The reaction is exothermic, so ΔH is negative, and the enthalpy change is $-1250\,kJ\,mol^{-1}$

Now test yourself

Tested

1 In an experiment, 0.95 g of powdered zinc (A_r = 65.4) is added to an excess of 25.0 cm^3 of 2.00 mol dm^{-3} hydrochloric acid. The temperature rose by 11.4°C. Calculate the enthalpy change, in kJ mol^{-1} for this reaction. The specific heat capacity of water is 4.18 J K^{-1} g^{-1}.

2 In an experiment, 0.16 g of methanol (CH_3OH, M_r = 32.0) is combusted when a spirit burner heats 100 cm^3 of water from 17.0°C to 25.0°C. Calculate the enthalpy change of combustion of methanol.

Answers on p. 105

Simple applications of Hess's Law

Hess's Law: the enthalpy change of a process is independent of the route taken, whether it be direct or indirect.

Enthalpy of formation and combustion

Revised

Standard enthalpy of formation, $\Delta H^\ominus_f$, is defined as the heat (or enthalpy) change when **1 mole** of a compound is formed from its **constituent elements, all reactants and products** in their **standard states** at 298 K and 100 kPa pressure.

The standard enthalpy of formation, $\Delta H^\ominus_f$, for any *element* in its standard state is, by definition, zero.

Equations to represent the enthalpy of formation of sodium chloride and of ethane gas in which one mole of the compound is formed are as follows:

$Na(s) + \frac{1}{2}Cl_2(g) \rightarrow NaCl(s)$

$2C(s) + 3H_2(g) \rightarrow C_2H_6(g)$

Standard enthalpy of combustion, $\Delta H^\ominus_c$, is defined as the enthalpy change when 1 mole of a substance is completely burned in oxygen under standard conditions, 298 K and 100 kPa pressure, with all reactants and products in their standard states.

For example, the equation that represents the enthalpy of combustion of liquid methanol ($CH_3OH(l)$) is:

$CH_3OH(l) + 1\frac{1}{2}O_2(g) \rightarrow CO_2(g) + 2H_2O(l)$

Determining unknown enthalpies

Revised

Cycle 1: Using enthalpies of formation

Consider the following reaction, for which the enthalpy change is unknown:

$4NH_3(g) + 5O_2(g) \rightarrow 4NO(g) + 6H_2O(l)$

$\Delta H^\ominus_f$ values (in kJ mol^{-1}) are: NO(g) = +90.4, $H_2O(l)$ = −286 and $NH_3(g)$ = −46.2.

Applying Hess's Law, the enthalpy change, $\Delta H^\ominus$, for this reaction can be determined by drawing an energy cycle using the $\Delta H^\ominus_f$ values as in Figure 6.2:

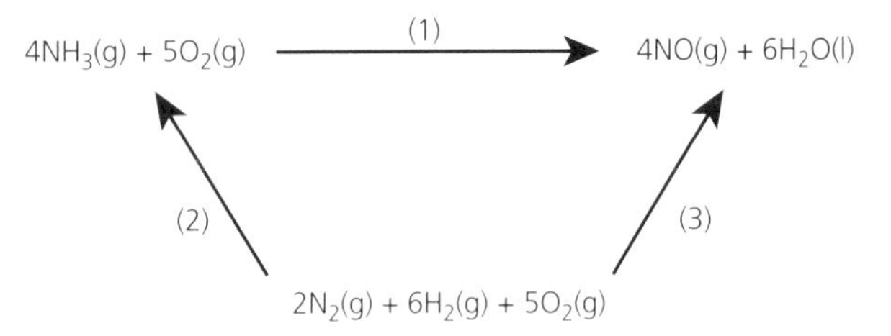

Figure 6.2

Using the cycle, it can be seen that enthalpy change (1) = −(2) + (3) = (3) − (2).

Letting $\Delta H^{\ominus}$ represent (1):

$$\Delta H^{\ominus} = 4\Delta H^{\ominus}_f[NO(g)] + 6\Delta H^{\ominus}_f[H_2O(l)] - 4\Delta H^{\ominus}_f[NH_3(g)] - 5\Delta H^{\ominus}_f[O_2(g)]$$

Substituting the appropriate given enthalpy of formation data gives:

$$\Delta H^{\ominus} = (4 \times 90.4) + (6 \times -286) - (4 \times -46.2) - (5 \times 0) = -1170\,kJ\,mol^{-1}.$$

Examiners' tip

It is sometimes more convenient to remember that if a question asks for $\Delta H^{\ominus}$ to be determined and only $\Delta H^{\ominus}_f$ data is provided then $\Delta H^{\ominus} = \Sigma\Delta H^{\ominus}_f$ (products) − $\Sigma\Delta H^{\ominus}_f$ (reactants).

Write balanced equations that represent the standard enthalpies of formation of these substances:

(a) carbon dioxide, $CO_2(g)$

(b) hexane, $C_6H_{14}(l)$

Calculate the enthalpy change for the reaction:

$$P_4O_{10}(s) + 6H_2O(l) \rightarrow 4H_3PO_4(l)$$

given the following standard enthalpies of formation (in $kJ\,mol^{-1}$): $P_4O_{10}(s) = -2984.0$, $H_2O(l) = -286$, $H_3PO_4(l) = -1279$.

Answers on p. 105

Cycle 2: Using enthalpies of combustion

Determine $\Delta H^{\ominus}_f$ for methane given that the enthalpies of combustion of methane, carbon and hydrogen are −890.4, −393.5 and −285.8 $kJ\,mol^{-1}$ respectively.

Answer

The equation representing the enthalpy of formation of methane is: $C(s) + 2H_2(g) \rightarrow CH_4(g)$

Figure 6.3 shows a cycle that has this equation across the top, and each combustion down the other sides:

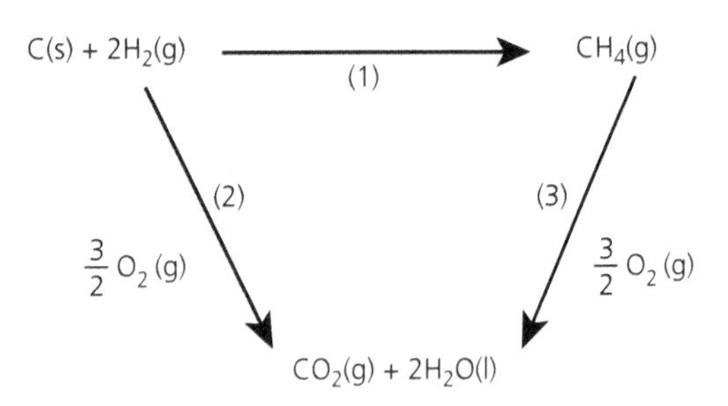

Figure 6.3

Enthalpy change for the direct route from elements to oxides = (2)

Enthalpy change for the indirect route from elements to oxides = (1) + (3)

Hence, (2) = (1) + (3) or (1) = (2) − (3)

$$\Delta H^{\ominus}_f[CH_4(g)] = \Delta H^{\ominus}_c[C(s)] + 2\Delta H^{\ominus}_c[H_2(g)] - \Delta H^{\ominus}_c[CH_4(g)]$$

$$= -393.5 + (2 \times -285.8) - (-890.4) = -74.7\,kJ\,mol^{-1}.$$

Now test yourself | Tested

5 Use the data below to calculate the enthalpy of formation of propane.

$C(s) + O_2(g) \rightarrow CO_2(g)$ $\Delta H = -393.5\,kJ\,mol^{-1}$

$H_2(g) + \frac{1}{2}O_2(g) \rightarrow H_2O(l)$ $\Delta H = -285.8\,kJ\,mol^{-1}$

$C_3H_8(g) + 5O_2(g) \rightarrow 3CO_2(g) + 4H_2O(l)$ $\Delta H = -2220.0\,kJ\,mol^{-1}$

Answers on p. 105

Bond enthalpies

Mean bond enthalpy

Revised

Mean bond enthalpy is defined as the average energy required to break **1 mole** of specified bonds in the **gas phase** measured under **standard conditions** (298 K and 100 kPa pressure) over several different compounds containing the bond of interest.

- The average bond enthalpy of the N–H bond is $+388\,kJ\,mol^{-1}$. This means that 388 kJ of heat energy are required to break 1 mole of N–H bonds in the gas phase. The decomposition of 1 mole of ammonia to form its constituent gaseous atoms:

  $NH_3(g) \rightarrow N(g) + 3H(g)$

  would require 3 × 388 kJ of heat energy, that is 1164 kJ, because three N–H bonds are being broken.
- For a hydrogen molecule, the change $H–H(g) \rightarrow 2H(g)$ represents the bond enthalpy of the H–H bond; this has a value of $+436\,kJ\,mol^{-1}$. However, unlike the H–H bond, many other covalent bonds (for example C–H, C–C, C=O and N–H) vary in their strength depending on the molecule in which the bond is present. For this reason, *mean* bond enthalpies are quoted for covalent bonds.
- Bond enthalpies are always endothermic because energy must be provided to overcome the electrostatic attraction between the shared electrons in the covalent bond and the positively charged nuclei.

Examiners' tip

Experimental enthalpy values may differ from those calculated using mean bond enthalpies. This is because mean bond enthalpies are averages and not compound-specific values.

Example

Calculate $\Delta H^{\ominus}$ for the following reaction using mean bond enthalpies:

$$CH_4(g) + 2Br_2(g) \rightarrow CH_2Br_2(g) + 2HBr(g)$$

Answer

The mean bond enthalpies, in $kJ\,mol^{-1}$, are C–H = +412, Br–Br = +193, C–Br = +276, H–Br = +366.

The energy cycle shown in Figure 6.4 can be drawn to represent the reaction.

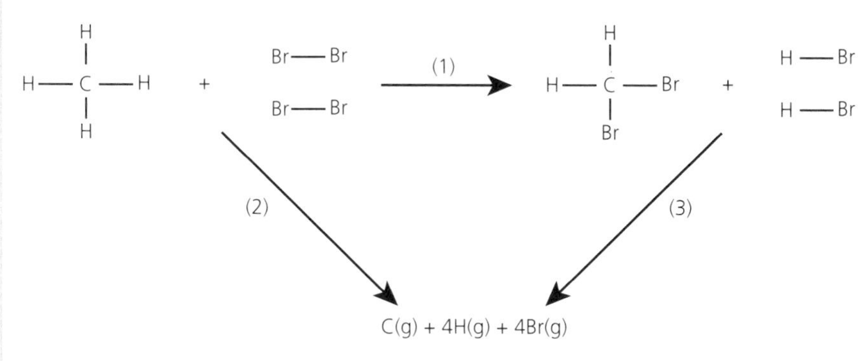

Figure 6.4

From the cycle, it can be seen that (1) = (2) − (3).

Using E to represent mean bond enthalpy:

$\Delta H(1) = (4 \times E[\text{C–H}]) + (2 \times E[\text{Br–Br}]) - (2 \times E[\text{C–H}]) - (2 \times E[\text{C–Br}]) - (2 \times E[\text{H–Br}])$

Substituting the relevant bond enthalpies gives:

$(4 \times 412) + (2 \times 193) - (2 \times 412) - (2 \times 276) - (2 \times 366) = -74\ \text{kJ mol}^{-1}$.

Examiners' tip

In bond energy calculations, it is sometimes easier to remember:

ΔH = bond-breaking energy − bond-making energy

or simply 'break minus make'.

The mean bond enthalpies N=N = 409, H–H = 436, N–H = 388 and N–N = 163 are in kJ mol^{-1}.
Calculate the approximate enthalpy change, ΔH, for this reaction of hydrogen:

$N_2H_2(g) + H_2(g) \rightarrow N_2H_4(g)$

```
                              H       H
N = N       H                  \     /
/     \     |      ------>      N — N
H       H   H                  /     \
                              H       H
```

Answers on p. 105

Exam practice

1 The aim of this question is to determine the enthalpy change for the reaction between 0.56 g of calcium (A_r = 40.1) and 25.0 cm^3 of 2.0 mol dm^{-3} dilute hydrochloric acid. The initial temperature of the hydrochloric acid is 21.1°C and it reaches a final temperature of 32.0°C. The specific heat capacity of water is 4.18 $\text{J K}^{-1}\text{g}^{-1}$

(a) Calculate the number of moles of calcium used. [1]

(b) Calculate the number of moles of acid used. [1]

(c) Write a balanced equation for the reaction. [1]

(d) Hence show that the acid is in excess in the reaction. [1]

(e) Calculate a value for the amount of heat evolved in this experiment in kJ. [2]

(f) Convert your answer in part (e) into a value for the enthalpy change, ΔH, in kJ mol^{-1}. [1]

2 For the reaction: $I_2(g) + Cl_2(g) \rightarrow 2ICl(g)\ \Delta H = -11\ \text{kJ mol}^{-1}$.

If the mean bond enthalpies are E(I–I) = +158 kJ mol^{-1} and Cl–Cl = +242 kJ mol^{-1}, calculate the I–Cl bond enthalpy. [3]

3 Calculate the enthalpy of combustion of but-1-ene, $C_4H_8(g)$, given that the enthalpy of formation of but-1-ene, carbon dioxide and water are +1.2, −393.5 and −285.8 kJ mol^{-1} respectively. [3]

Answers and quick quiz 4 online

Online

Examiners' summary

You should now have an understanding of:

- ✓ what is meant by 'enthalpy change'
- ✓ how to carry out simple experiments to enable enthalpy changes to be measured
- ✓ Hess's law
- ✓ enthalpy of formation
- ✓ enthalpy of combustion
- ✓ how to apply Hess's law by drawing simple energy cycles
- ✓ mean bond enthalpies and using them to calculate enthalpy changes

7 Kinetics

Rate of reaction is defined as the **rate of change of concentration with time**. It can expressed mathematically in the form:

$$\text{rate of reaction (mol dm}^{-3}\text{s}^{-1}) = \frac{\text{change in concentration (mol dm}^{-3})}{\text{time for concentration change (s)}}$$

Collision theory

Effective collisions

Revised

For reactions to occur, particles with sufficient energy must collide — the type of collision that leads to a reaction is called an *effective* collision. If particles do not have enough energy for a reaction, or do not collide with the correct alignment or orientation, an ineffective or inelastic collision may result.

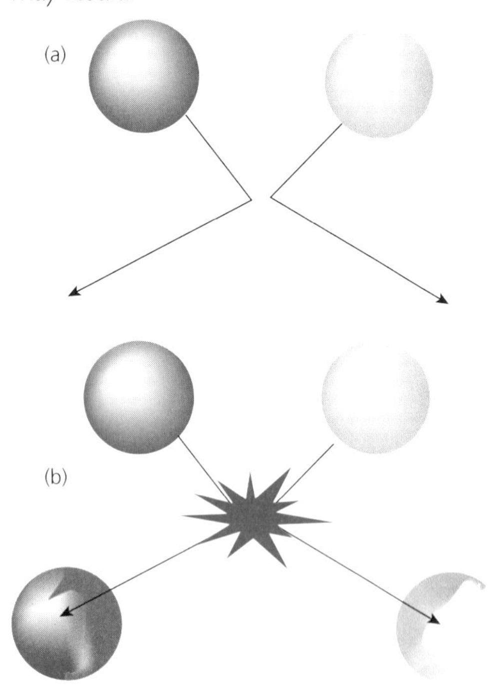

Figure 7.1 Collisions in reactions: (a) ineffective; (b) effective

The **activation energy, E_a**, is the minimum energy required for a reaction to take place. If the combined energy of the particles colliding is less than the activation energy, then a reaction is not likely to occur — this is an **ineffective collision**. If the combined energy of the particles exceeds the activation energy, then an **effective collision** is more likely.

Most collisions that take place between particles do not lead to a reaction because the combined energy of the particles is lower than the activation energy.

Factors affecting rate of reaction

Changing the temperature

Revised

As the temperature increases, so too does the average kinetic energy of the particles in a reaction mixture. Particles will collide with a greater combined energy and if this exceeds the reaction activation energy then a reaction may occur.

The distribution of molecules at two different temperatures, T_1 and T_2, can be represented on a **Maxwell–Boltzmann distribution** such as shown in Figure 7.2.

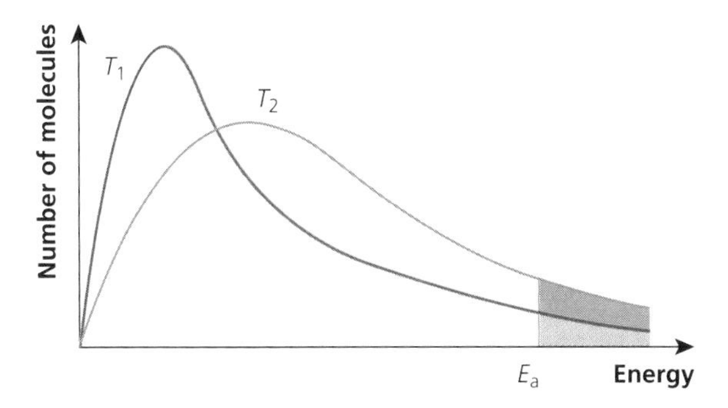

Figure 7.2 Maxwell–Boltzmann distributions at different temperatures

- At the lower temperature, T_1, the proportion of molecules with an energy greater than the activation energy, E_a, is given by the dark-shaded area.
- If the temperature is increased to T_2, a new distribution is formed in which the peak of the graph shifts to higher energy. A greater proportion of particles possess an energy greater than the activation energy — the light-shaded area — and therefore there will be more successful collisions at the higher temperature.
- Rate of reaction is normally very sensitive to temperature increases. This is because the proportion of molecules having an energy exceeding the activation energy — the area to the right of E_a — increases dramatically (exponentially) as the temperature increases. A small temperature increase can cause a significant shift of the Maxwell–Boltzmann distribution to the right, therefore moving more molecules to higher energies.

Typical mistake

Many candidates state that it is the fact that molecules collide more often at a higher temperature that causes the greater rate of reaction. It is because more molecules are colliding with greater **energy**, and therefore with sufficient energy to react, that has the much greater effect.

Now test yourself

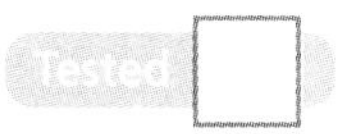

1 The rate of reaction between hydrogen gas and oxygen gas to form water is very sensitive to changes in temperature:

$$2H_2(g) + O_2(g) \rightarrow 2H_2O(l)$$

Suggest why an increase in temperature increases the rate of this reaction.

Answers on p. 105

Changing the concentration

Revised

Concentration is a measure of the number of particles per unit volume and is normally expressed in $mol\,dm^{-3}$. The greater the concentration, the higher the number of particles per unit volume.

As the concentration of a solution increases, there will be more particles in the same volume. More collisions per unit time will occur and a greater number of effective collisions will take place, giving a greater rate of reaction.

In the left-hand box in Figure 7.3, the concentration of red particles is much lower than that in the right-hand box. This means that fewer collisions will take place per unit time between the red and blue balls on the left-hand side than on the right. This would give a lower rate of reaction because fewer effective collisions would result.

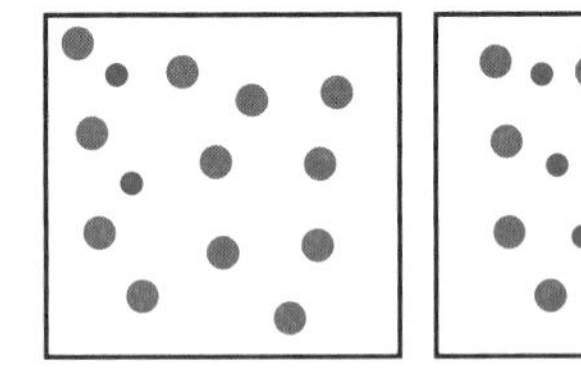

Figure 7.3 The effect of concentration

Using a catalyst

Revised

A **catalyst** is a substance that increases the rate of a reaction by **providing an alternative reaction pathway** with a **lower activation energy**. At the end of the reaction a catalyst is chemically unchanged.

Manganese(IV) oxide is a **catalyst** in the decomposition of hydrogen peroxide:

$$2H_2O_2(aq) \rightarrow 2H_2O(l) + O_2(g)$$

Without the catalyst, the reaction rate would be significantly slower. When manganese(IV) oxide is added, it provides an alternative reaction pathway, or mechanism, by which the reaction can take place. This alternative route is faster, because it has a lower activation energy.

If the activation energy is lowered (see Figure 7.4) there will be a greater proportion of molecules with an energy exceeding the old activation energy threshold and a faster reaction rate results.

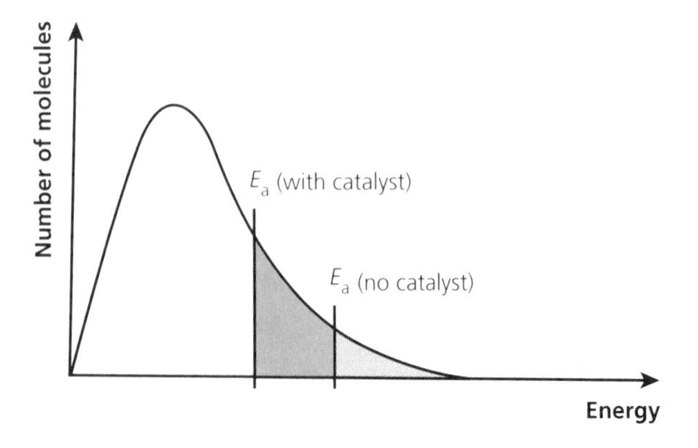

Figure 7.4 The effect of a catalyst on activation energy

Examiners' tip

Note that the activation energy is lowered using a catalyst. However, when explaining the effect of temperature and concentration on rate, the activation energy stays the same.

You can also sketch an energy profile for a reaction to show how a catalyst affects the activation energy for a process, as shown in Figure 7.5. Note the following about this diagram:

- the activation energy for the uncatalysed reaction is much larger than that for the catalysed reaction
- the enthalpy change, ΔH, for the reaction is unchanged — a catalyst affects the rate but not the amount of heat absorbed or released.

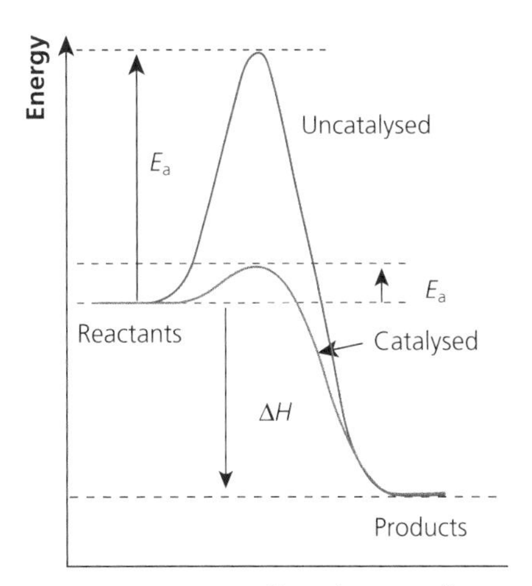

Figure 7.5 Energy profile for a catalysed reaction

Now test yourself

2 The reaction between sodium thiosulfate solution, $Na_2S_2O_3(aq)$, and dilute hydrochloric acid can be represented by:

$$Na_2S_2O_3(aq) + 2HCl(aq) \rightarrow 2NaCl(aq) + S(s) + SO_2(g) + H_2O(l)$$

Explain why the rate of this reaction increases when the concentration of sodium thiosulfate increases.

3 Give an example of a catalyst in use, and explain how it increases the rate of the chemical reaction.

Answers on p. 105

Tested

Exam practice

1 **(a)** Give the meaning of the term 'activation energy'. [1]

Figure 7.6 shows a distribution of molecular energies in a mixture of gases at a particular temperature. The activation energy is also shown for a reaction between the gases.

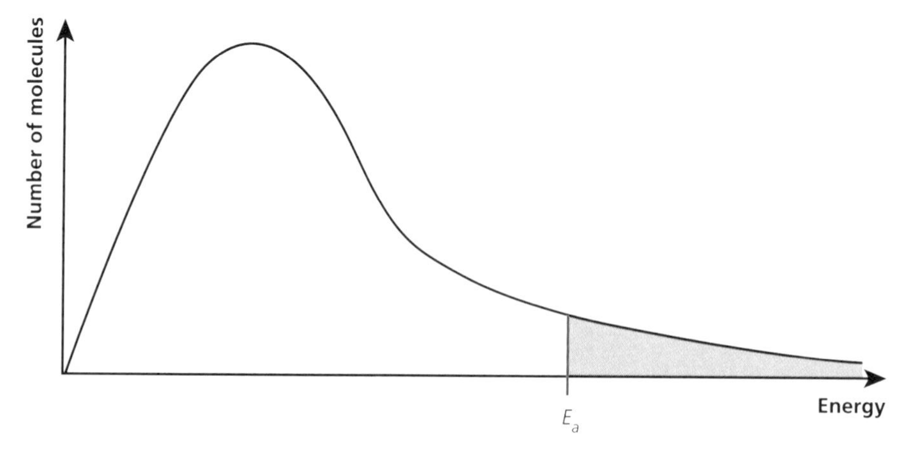

Figure 7.6

(b) Label on the graph:

the most common energy [1]

molecules with the highest energy [1]

molecules with the lowest energy [1]

(c) On the same axes, draw the graph expected when the reaction mixture is heated by a further 10°C. [2]

(d) Use your graphs to explain why an increase in temperature of 10°C results in an average increase in the rate of reaction by a factor of approximately 2. [2]

2 For the reaction $2HI(g) \rightarrow H_2(g) + I_2(g)$, the activation energy when uncatalysed is 183 kJ mol^{-1} and when catalysed with gold it is 105 kJ mol^{-1}. ΔH for the reaction is −52 kJ mol^{-1}.

(a) Sketch an energy profile diagram for the reaction, including the activation energies for both the uncatalysed and catalysed reactions. [2]

(b) Calculate the activation energy for the *reverse reaction* [$H_2(g) + I_2(g) \rightarrow 2HI(g)$] in both the uncatalysed and catalysed reactions. [2]

(c) Explain why increasing the concentration of hydrogen iodide gas results in a faster reaction rate. [2]

Answers and quick quiz 4 online

Online

Examiners' summary

You should now have an understanding of:

- collision theory — particles have to collide to react
- particles need to possess a certain minimum energy to react
- activation energy
- how temperature effects the rate of reaction
- the Maxwell–Boltzmann distribution
- how concentration affects the rate of reaction
- what is meant by 'catalysts' and how they work

8 Equilibria

The dynamic nature of equilibria

Reversible reactions

Revised

There are many reactions that can be described as being 'reversible' — can proceed in either the forward or reverse directions. For example, ethene can be hydrated to form ethanol, C_2H_5OH

$$C_2H_4(g) + H_2O(g) \rightarrow C_2H_5OH(g)$$

On the other hand, ethanol can be dehydrated to form ethene and water:

$$C_2H_5OH(l) \rightarrow C_2H_4(g) + H_2O(g)$$

When a reaction that is reversible takes place, both the forward and reverse processes will occur until a **state of dynamic chemical equilibrium** is attained. When the system (reaction) reaches this stage, the rates of the forward and reverse reactions are equal.

Examiners' tip

When a reaction is at **equilibrium**, the concentrations of reactants and products are not equal, but the **rates** of the forward and reverse processes are equal.

When at equilibrium ($\rightleftharpoons$) the reaction above can be represented as:

$$C_2H_4(g) + H_2O(g) \rightleftharpoons C_2H_5OH(g)$$

Another example of making an alcohol in an industrial process is when carbon monoxide reacts with hydrogen to form methanol, CH_3OH

$$CO(g) + 2H_2(g) \rightleftharpoons CH_3OH(g)$$

A state of equilibrium will occur only if the system is **closed**, so that nothing is allowed to enter or leave the reaction.

Examiners' tip

Both ethanol and methanol are important fuels, and both are formed industrially in equilibrium processes.

Qualitative effects of changes in external conditions

Equilibrium position

Revised

- Reactions that come to equilibrium will consist of varying relative proportions of reactants and products. If a reaction comes to equilibrium and the quantity of reactants is greater than the quantity of products, the reaction is said to have an equilibrium position that lies on the **left-hand side**.
- The **equilibrium position** therefore gives a measure of the extent to which the reaction takes place, and the position of equilibrium may be 'shifted' by changing the external conditions.

Le Chatelier's principle

Revised

Using **Le Chatelier's principle**, it is possible to predict the qualitative effect of the changing external conditions — for example concentration, pressure or temperature — on an equilibrium position.

Le Chatelier's principle: 'If the conditions under which an equilibrium exists are changed, the position of equilibrium alters in such a way as to oppose the change in conditions.'

Changing the concentration of a reactant or a product

If the concentration of a substance changes, the reaction shifts position so as to oppose the concentration change.

Consider this example. The following process involving the formation of the ester methyl methanoate ($HCOOCH_3$) has reached equilibrium:

$$CH_3OH(l) + HCOOH(l) \rightleftharpoons HCOOCH_3(l) + H_2O(l)$$

- If the concentration of methanol, CH_3OH, is increased by adding more moles of CH_3OH, the position of equilibrium position moves to lower the concentration of the CH_3OH — it shifts to the **right-hand side** to produce more ester.
- If methanoic acid, HCOOH, is removed from the equilibrium by reducing its concentration, the equilibrium position shifts so as to produce more HCOOH — the equilibrium shifts to the **left-hand side**.

Changing the pressure

If a reaction involves gases, the equilibrium position may be affected by changes in external pressure. Consider the manufacture of ammonia:

$$N_2(g) + 3H_2(g) \rightleftharpoons 2NH_3(g)$$

- There is a total of 4 molecules of gas on the left-hand side of the equilibrium and only 2 molecules of gas on the right-hand side, therefore the left-hand side (the reactants) will exert the greater pressure relative to the products.
- If the external pressure is increased, the equilibrium position moves so as to decrease the pressure. The reaction will do this by shifting to the right-hand side to produce more molecules of ammonia and fewer molecules of nitrogen and hydrogen, i.e. fewer gas molecules overall.

However, if the reaction being considered is:

$$H_2(g) + I_2(g) \rightleftharpoons 2HI(g)$$

- A change in external pressure will not affect the equilibrium position because there are equal numbers of moles of gas on each side of the equation.
- The *rate* at which equilibrium is attained will be affected because pressure affects the relative distance between molecules in the gas phase. The collision rate will therefore change and the number of effective collisions will also change.

Changing the temperature

To predict how a change of temperature affects the equilibrium position, the sign of the enthalpy change, ΔH, for the reaction has to be known. In the reaction:

$$N_2(g) + O_2(g) \rightleftharpoons 2NO(g);\ \Delta H = +180.5\,kJ\,mol^{-1}$$

- The **forward** reaction is **endothermic**. Therefore formation of NO(g) is accompanied by the removal of heat from the surroundings.
- The **reverse** process is **exothermic**. Heat will be released when $N_2(g)$ and $O_2(g)$ are formed.
- If the external temperature is increased, the reaction will try to remove excess heat by using its forward process and more NO(g) will form.

- Temperature also affects the *rates* of both the forward and the reverse process (but not equally). If the temperature is increased, the rates of both processes increase and the rate at which the equilibrium is attained also increases.

Catalysts

Revised

A **catalyst** will not affect the equilibrium *position*, it will only affect the *rate* at which the equilibrium is attained.

The ester ethyl ethanoate, $CH_3COOC_2H_5$, is formed in this reaction:

$C_2H_5OH(l) + CH_3COOH(l) \rightleftharpoons CH_3COOC_2H_5(l) + H_2O(l)$

- It can take a long time for this reaction to come to equilibrium — many months, or even years.
- If a catalyst such as concentrated sulfuric acid is added, the equilibrium is attained faster.
- The rates of the forward and reverse processes are increased equally by the catalyst.

The importance of equilibria in industrial processes

Ethanol production

Revised

One very important example of a reversible reaction that comes to a state of dynamic equilibrium is the industrial manufacture of ethanol from steam and ethene:

$C_2H_4(g) + H_2O(g) \rightleftharpoons C_2H_5OH(g)$ $\quad \Delta H$ is exothermic

Ethene reacts with steam in the presence of a phosphoric acid catalyst (on a silica support) at 300°C, and a pressure of about 60–70 atmospheres.

The process is reversible and so it is important to appreciate how conditions are changed so that a greater yield of ethanol is formed at an appreciable rate.

Pressure

- **Yield:** in the equation for the process, there are fewer gas molecules (low pressure) on the product side, and more gas molecules (high pressure) on the reactant side. When the external pressure is increased, the equilibrium will shift to the right-hand side to reduce the pressure, so a greater yield of ethanol will form.
- **Rate:** when gases are compressed, there will be more collisions per unit time because the molecules will be closer together. The rate at which equilibrium is achieved therefore increases.

Temperature

- **Yield:** decreasing the temperature will increase the yield of ethanol, because the forward exothermic reaction will be favoured.
- **Rate:** a lower temperature will result in a lower rate of reaction.

A compromise is found between yield and rate by adopting a moderate temperature.

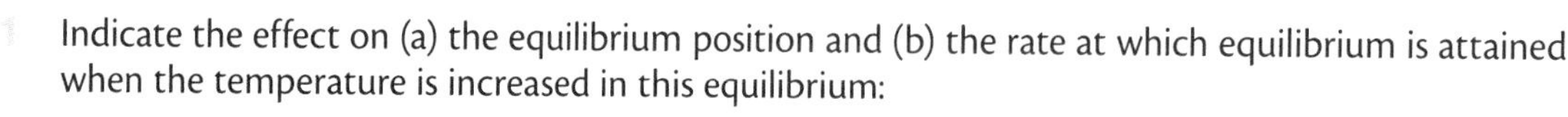

1 Indicate the effect on (a) the equilibrium position and (b) the rate at which equilibrium is attained when the temperature is increased in this equilibrium:

$2NO_2(g) \rightleftharpoons N_2O_4(g)$

The forward reaction is exothermic.

Examiners' tip

Yield and rate are different features of an equilibrium system — Le Chatelier's principle can be used to explain the changes to the yield; collision theory explains the rate changes.

Catalyst

The phosphoric acid catalyst affects the rate at which the equilibrium is attained; it does not affect the yield of ethanol.

2 Copy and complete the table below to describe some changes to the equilibrium:

$CO(g) + 2H_2(g) \rightleftharpoons CH_3OH(g)$

The forward reaction is exothermic.

	Equilibrium position	Rate at which equilibrium is attained
Total pressure is increased		
Temperature is increased		
A catalyst is added		

Exam practice

1 In the Haber Process, nitrogen and hydrogen react in the presence of an iron catalyst to form ammonia according to the equation:

$N_2(g) + 3H_2(g) \rightleftharpoons 2NH_3(g)$

(a) The production of ammonia is accompanied by heat being released. Explain whether a low temperature or a high temperature would result in the higher yield of ammonia. [2]

(b) State what happens to the rate at which equilibrium is attained when:

(i) the temperature is increased [1]

(ii) the pressure is decreased [1]

(c) Explain your answers to part (b) in terms of energies of particles. [2]

(d) Explain why a moderately high temperature of 450°C is used when manufacturing ammonia industrially. [2]

(e) Explain why pressures of more than 200–350 atmospheres are rarely used in the Haber process. [2]

(f) In one particular year, the United States of America produced 8 million tonnes of ammonia at a cost of $200 per tonne.

(i) Calculate the total cost of the U.S. production of ammonia. [1]

(ii) Calculate the mass of nitrogen gas required to produce 8 million tonnes of ammonia.

[A_r data: N = 14.0, H = 1.0] [2]

(g) Given that the process produces an overall ammonia yield of 15%, calculate the mass of nitrogen required when this yield is taken into account. [1]

Answers and quick quiz 4 online

Online

Examiners' summary

You should now have an understanding of:

- ✓ reversible reactions
- ✓ what is meant by 'state of equilibrium'
- ✓ what is meant by 'dynamic equilibrium'
- ✓ how Le Chatelier's principle can be used to predict the direction in which an equilibrium shifts when external conditions are changed
- ✓ the production of ethanol in terms of how to change external conditions to optimise both yield and rate

9 Redox reactions

Oxidation and reduction

Redox reaction

Revised

In a **redox reaction**, one substance is oxidised and another is reduced. The substance that oxidises the other is called an **oxidising agent**. The substance that reduces the other is called the **reducing agent**.

Oxidation is the loss of electrons.
Reduction is the gain of electrons.

A simple redox reaction takes place when a metal compound, like aluminium oxide, is electrolysed. These processes take place at the electrodes:

- at the (–) cathode: $Al^{3+}(l) + 3e^- \rightarrow Al(l)$
- at the (+) anode: $2O^{2-}(l) \rightarrow O_2(g) + 4e^-$

Aluminium ions gain electrons, so they are **reduced**.

Oxide ions lose electrons, so they are **oxidised**.

Oxidation states

Working out the oxidation state

Revised

An oxidation state is a number indicating the 'formal' charge that an element would have in a compound if the compound were ionic. The oxidation state of an element is zero.

Working out the oxidation state for a combined element in a compound assumes that all the other oxidation states are known. The sum of all of the oxidation states in a compound is zero.

Some typical oxidation states:

- hydrogen (except when in hydrides) = +1
- group 1 metals = +1
- group 2 metals = +2
- oxygen (except when in hydrogen peroxide) = –2
- fluorine = –1

Example

Work out the oxidation state of the following:

(a) iron in Fe_2O_3

(b) manganese in $KMnO_4$

(c) chromium in $Na_2Cr_2O_7$

Answer

(a) In Fe_2O_3 the known oxidation state is oxygen at −2
Therefore 2Fe + 3(−2) = 0
2Fe = +6, so each iron is +3
This compound is called iron(III) oxide, where 'III' is the oxidation state of the iron in the compound.

(b) In $KMnO_4$ the known oxidation states are potassium at +1 and oxygen at −2
Therefore (+1) + Mn + 4(−2) = 0
Mn = (+8) − 1, so each manganese is +7
The compound is called potassium manganate(VII), where 'VII' is the oxidation state of manganese in the compound.

(c) In $Na_2Cr_2O_7$ the known oxidation states are sodium at +1 and oxygen at −2
Therefore, 2(+1) + 2Cr + 7(−2) = 0
2 + 2Cr − 14 = 0, so 2Cr = 14 − 2
Cr = +6, so each chromium is +6
This compound is called sodium dichromate(VI), where 'VI' is the oxidation state of chromium in the compound.

For complex ions such as SO_4^{2-}, the sum of the oxidation states is equal to the charge on the ion.

Example

What is the oxidation state of aluminium in the ion $[AlF_6]^{3-}$?

Answer

Known oxidation states: F = −1

Therefore, Al + 6(−1) = −3

Al − 6 = −3

So Al = −3 + 6 = +3

Now test yourself Tested ☐

1 What is the oxidation state of the named element in each of these compounds and ions?
   (a) cobalt in $CoCl_3$
   (b) chlorine in NaOCl
   (c) titanium in $TiCl_4$
   (d) iron in Na_2FeO_4
   (e) sulfur in H_2SO_4
   (f) iodine in IO_3^-
   (g) manganese in MnO_4^{2-}

2 Name, using oxidation states, the compounds in question 1.

Answers on p. 105

Using oxidation states

Revised

When carrying out a reaction, it is possible to deduce whether the reaction is redox or not, and if it is, which elements have been oxidised and which reduced.

Examiners' tip

If the oxidation state of an element increases this is an **oxidation**; if it decreases this is a **reduction**.

Determine whether the following are redox reactions or not by deducing the oxidation state changes.

(a) $H_2SO_4(l) + NaCl(s) \rightarrow NaHSO_4(s) + HCl(g)$

(b) $2SO_2(g) + O_2(g) + 2H_2O(l) \rightarrow 2H_2SO_4(aq)$

Answer

(a) For H_2SO_4: hydrogen = +1; oxygen = −2, so sulfur = +6
NaCl: sodium = +1; chlorine = −1
$NaHSO_4$: sodium = +1; hydrogen = +1; oxygen = −2, so sulfur = +6
HCl: hydrogen = +1; chlorine = −1
So all the oxidation states before and after the reaction are the same, so this is not a redox reaction.

(b) For SO_2: sulfur is +4; oxygen is −2
O_2: oxygen is 0
H_2O: hydrogen is +1 and oxygen is −2
H_2SO_4: hydrogen is +1, sulfur is +6 and oxygen is −2.
Hydrogen's oxidation state stays the same.
Sulfur changes oxidation state from +4 to +6. It has been oxidised.
Oxygen (starting in O_2) changes from 0 to −2. It has been reduced.
It is therefore a redox reaction.

Deduce the names of the oxidising and reducing agents in the reaction:

$2SO_2(g) + O_2(g) + 2H_2O(l) \rightarrow 2H_2SO_4(aq)$

Show that the reaction below is a redox reaction by determining the oxidation states of the elements that change. Also deduce the names of the oxidising and reducing agents:

$PbO_2(s) + 4HCl(aq) \rightarrow PbCl_2(aq) + 2H_2O(l) + 2Cl_2(g)$

Redox equations

Writing half-equations

Revised

Redox reactions consist of two separate processes — an oxidation and a reduction. These can be written separately in the form of **half-equations**.

Example

Magnesium metal is added to copper(II) sulfate solution to form copper and magnesium sulfate solution. Write a balanced symbol equation for the reaction and also the half-equations. Then determine which element has been oxidised and which has been reduced.

Answer

$$Mg(s) + CuSO_4(aq) \rightarrow MgSO_4(aq) + Cu(s)$$

- In this reaction, the magnesium atoms lose two electrons to form magnesium ions, Mg^{2+}.
- Half-equation: $Mg(s) \rightarrow Mg^{2+}(aq) + 2e^-$; the magnesium atoms have therefore been oxidised.
- The copper(II) ions , Cu^{2+}, gain two electrons to form copper atoms, Cu.
- Half-equation: $Cu^{2+}(aq) + 2e^- \rightarrow Cu(s)$; the copper(II) ions have therefore been reduced.

Now test yourself

Tested

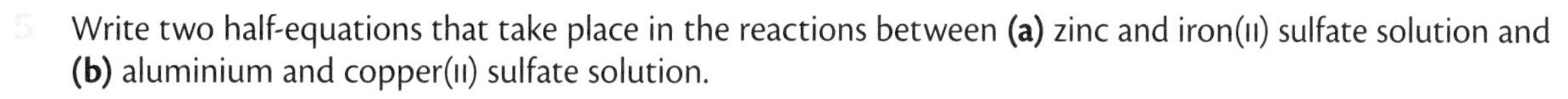

5 Write two half-equations that take place in the reactions between **(a)** zinc and iron(II) sulfate solution and **(b)** aluminium and copper(II) sulfate solution.

Answers on p. 105

Two half-equations can be combined to form an overall **ionic equation**:

$$Cu^{2+}(aq) + Mg(s) \rightarrow Mg^{2+}(aq) + Cu(s)$$

When oxidation and reduction half-equations are combined, the electrons **must** cancel out.

In this reaction, magnesium has been oxidised and the copper(II) ion is the **oxidising agent** because it caused this oxidation. Copper(II) ions are reduced and so magnesium is called the **reducing agent**.

Example 1

Given the following half-equations for a redox reaction between chlorine and bromide ions, combine them to produce an overall ionic equation:

$$Cl_2(aq) + 2e^- \rightarrow 2Cl^-(aq)$$

$$2Br^-(aq) \rightarrow Br_2(aq) + 2e^-$$

Answer

Adding these two together, and cancelling the electrons on each side, gives:

$$Cl_2(aq) + 2Br^-(aq) \rightarrow Br_2(aq) + 2Cl^-(aq)$$

Example 2

When lithium metal is added to water, reactions occur according to these two half-equations:

$$Li(s) \rightarrow Li^+(aq) + e^-$$

$$2H_2O(l) + 2e^- \rightarrow 2OH^-(aq) + H_2(g)$$

What is the balanced equation for the reaction?

Answer

Multiplying the first equation by 2 to balance the electrons, then adding and cancelling the electrons gives:

$$2Li + 2H_2O \rightarrow 2Li^+ + 2OH^- + H_2$$

So the required equation is $2Li(s) + 2H_2O(l) \rightarrow 2LiOH(aq) + H_2(g)$

Exam practice

1 Nitrogen dioxide reacts with water:

$$2NO_2(g) + H_2O(l) \rightarrow H^+(aq) + NO_3^-(aq) + HNO_2(aq)$$

(a) Give the oxidation states of nitrogen in all of the nitrogen-containing species in the equation. [3]

(b) Which substance has been oxidised in the reaction? [1]

2 Scrap iron can be used to extract copper from dilute aqueous solutions containing copper(II) ions. Write the simplest ionic equation for the reaction of iron with copper(II) ions in aqueous solution. [1]

3 Ammonia reacts with oxygen in the presence of a platinum catalyst to form nitrogen(II) oxide and steam.

(a) Write a balanced equation for the reaction. [2]

(b) Identify which substances have been reduced and oxidised in the reaction. [2]

(c) What is the name of the reducing agent in the process? [1]

Answers and quick quiz 5 online

Online

Examiners' summary

You should now have an understanding of:

- what is meant by oxidation and reduction in terms of electron transfer
- what is meant by a redox reaction
- oxidation state
- how to deduce the oxidation state of a combined element in a compound
- how to use oxidation states to determine which elements have been oxidised and reduced in a reaction
- half-equations and how to write them for simple reactions
- how to combine half-equations to give ionic equations for redox reactions

10 The periodic table

Group 7 elements, the halogens

- The halogens are fluorine (F), chlorine (Cl), bromine (Br), iodine (I) and astatine (At) — they are members of group 7 of the periodic table.
- All halogens exist as diatomic molecules at room temperature and pressure, and are often given the general symbol X_2

Trends in physical properties

Revised

Boiling point

Boiling points increase down the group.

- The number of electrons in each molecule is increasing.
- This increases the number and strength of the **van der Waals' interactions** between molecules.
- So it becomes harder to separate one molecule from another.

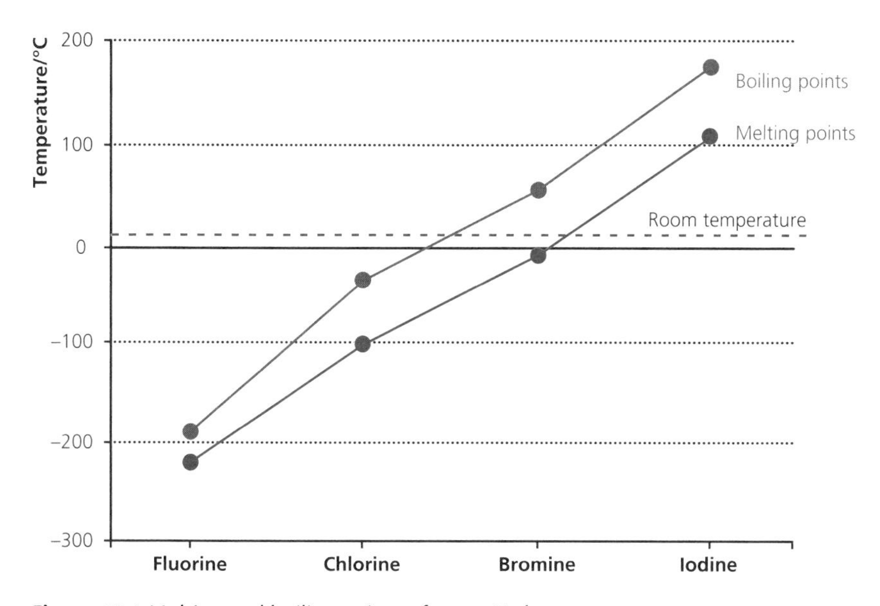

Figure 10.1 Melting and boiling points of group 7 elements

At room temperature fluorine is a pale-yellow gas, chlorine is a pale-green gas, bromine is an orange liquid and iodine is a dark-grey solid. Figure 10.1 shows clearly that the melting points and boiling points increase regularly down the group.

Electronegativity

Electronegativity **decreases** down the group.

- The atomic radius increases because of the added electron shells.
- Therefore the bonding pair of electrons in a covalent bond gets progressively further from the positively charged nucleus. There will also be additional shielding.

- This means that there will be less electrostatic attraction between the nucleus and the bonding electron pair as the group is descended, and the electronegativity decreases.

Examiners' tip

Remember that the term 'electronegativity' refers to the ability of an atom to attract pairs of **bonding** electrons in covalent bonds.

Trends in the oxidising ability of halogens

Revised

Halogen molecules, X_2, react by gaining electrons from other substances. A half-equation to show this process would be $X_2 + 2e^- \rightarrow 2X^-$.

In this process, the halogen molecule has been **reduced** because it has gained electrons. The other substance has been **oxidised** because it has had electrons removed by the halogen. This is why halogens are **oxidising agents**.

For example, sodium reacts with chlorine according to:

$$2Na(s) + Cl_2(g) \rightarrow 2NaCl(s)$$

Chlorine oxidises the sodium from oxidation state 0 to +1. Chlorine itself is reduced and changes oxidation state from 0 to −1.

The **oxidising power** of the halogens **decreases** down the group.

- As the group is descended, the number of electron energy levels increases.
- The distance between the outer electron shell and the nucleus (the atomic radius) increases.
- Shielding increases between the outer shell electrons and the nucleus.
- Because of the extra distance and shielding, the larger elements' atoms attract an extra electron less strongly.
- Hence the oxidising power decreases down the group, and also the ability of a halogen molecule to be reduced.

Halogen–halide displacement reactions

A more reactive halogen will displace a less reactive halogen from its compounds.

Examiners' tip

Make sure that you use the correct words — 'chlorine' for the molecular element and 'chloride' for the ion.

When chlorine gas, $Cl_2(g)$, is bubbled through a solution containing sodium bromide, NaBr(aq), a **redox reaction** occurs in which the chlorine **oxidises** the bromide ions to aqueous bromine, $Br_2(aq)$.

- The overall equation for the reaction is
  $2NaBr(aq) + Cl_2(g) \rightarrow 2NaCl(aq) + Br_2(aq)$
- The half-equation to show bromide ions being oxidised is
  $2Br^-(aq) \rightarrow Br_2(aq) + 2e^-$
- The half-equation to show chlorine molecules being reduced is
  $Cl_2(g) + 2e^- \rightarrow 2Cl^-(aq)$

The order of oxidising power of halogens: $F_2 > Cl_2 > Br_2 > I_2$ (ignoring astatine)

The order of reducing power of halide ions: $I^- > Br^- > Cl^- > F^-$

Fluorine is a powerful oxidising agent and can oxidise Cl^-, Br^- and I^-

Chlorine can oxidise Br^- and I^-. Bromine can only oxidise I^-

Explain why chlorine is a more powerful oxidising agent than iodine.

(a) Give an equation that shows how bromine reacts with sodium iodide solution.

(b) Which substance is oxidised and which is reduced in this reaction?

Trends in the reducing ability of halide ions

Revised

Halide ions increase in **reducing power** as the group is descended.

- There are more electron energy levels as the group is descended.
- The **ionic radius** of the halide ion increases.
- The amount of shielding between the outer electrons and the nucleus increases.
- There is a lower electrostatic attraction between the nucleus and the electrons.
- So it is easier to remove an electron from a halide ion as the group is descended.
- Therefore, **halide ions** are more readily oxidised as the group is descended — they become more powerful reducing agents.

The change in reducing ability of the halide ions can be demonstrated using a reaction between a halide salt and concentrated sulfuric acid.

Using NaCl(s):

sodium chloride	+	concentrated sulfuric acid	→	sodium hydrogen sulfate(VI)	+	hydrogen chloride gas
NaCl(s)	+	H_2SO_4(l)	→	$NaHSO_4$(s)	+	HCl(g)

Using NaBr(s), and ignoring the Na^+ spectator ion:

$Br^-(s) + H_2SO_4(l) \rightarrow HSO_4^-(s) + HBr(g)$

then: $2HBr(g) + H_2SO_4(l) \rightarrow Br_2(g) + SO_2(g) + 2H_2O(l)$

Using NaI(s):

$I^-(s) + H_2SO_4(l) \rightarrow HSO_4^-(s) + HI(g)$

then: $8HI(g) + H_2SO_4(l) \rightarrow 4I_2(g) + H_2S(g) + 4H_2O(l)$

- In the first stage, the halide ion is protonated by the concentrated sulfuric acid: $X^- + H^+ \rightarrow HX$.
- If a bromide or an iodide is used, a second stage occurs in which the sulfuric acid is reduced.
- The oxidation state of sulfur in sulfuric acid is +6. This is reduced to +4 in SO_2 and −2 in H_2S.
- The iodide ion is the most powerful reducing agent of the halide ions and produces a mixture of products including SO_2, S and H_2S in the reaction with sulfuric acid.

Typical mistake

Many candidates write an equation showing the formation of sodium sulfate(VI), Na_2SO_4. It is sodium **hydrogen**sulfate(VI) that forms when concentrated sulfuric acid is used.

Now test yourself

Tested

3 Write an equation to show how potassium fluoride reacts with concentrated sulfuric acid.

4 **(a)** What are the different oxidation states of sulfur in H_2S and SO_2, and S?

(b) Write an equation for the reaction between hydrogen iodide and concentrated sulfuric acid to form sulfur, iodine and water.

Answers on p. 105

Identification of halide ions using silver nitrate

Revised

Halide ions, X^-, can be identified using silver nitrate solution acidified with nitric acid.

Nitric acid is used in the halide ion test because it reacts with any carbonates that may be present before silver nitrate is added. Silver ions react with carbonate ions to form insoluable white silver carbonate, and this will confuse the test.

Using calcium chloride, addition of acidified silver nitrate solution produces a white precipitate of silver chloride:

$$CaCl_2(aq) + 2AgNO_3(aq) \rightarrow Ca(NO_3)_2(aq) + 2AgCl(s)$$

Write an ionic equation for the reaction.

Answer

The ionic equation is $Ag^+(aq) + Cl^-(aq) \rightarrow AgCl(s)$.

The silver halide precipitates have similar colours and can be difficult to tell apart. A further test using aqueous ammonia of different concentrations can distinguish between them.

Table 10.1 Distinguishing silver halides

Silver halide formed	Colour of precipitate	Solubility in dilute $NH_3(aq)$	Solubility in conc. $NH_3(aq)$
Silver chloride, AgCl(s)	White	Soluble — a colourless solution forms	Soluble — a colourless solution forms
Silver bromide, AgBr(s)	Cream	Insoluble	Soluble — a colourless solution forms
Silver iodide, AgI(s)	Yellow	Insoluble	Insoluble

A solution of a substance Y is acidified with nitric acid and then a few drops of aqueous silver nitrate are added. A cream precipitate is formed. The precipitate is insoluble in dilute ammonia solution but does dissolve in concentrated ammonia.

What can you deduce from this information?

Answer

You can deduce that bromide ions must have been present in the compound, because the precipitate was cream in colour and dissolved in concentrated ammonia, but not in dilute ammonia.

Revision activity

Prepare some flash cards — on one side you write the type of halogen reaction (for example 'displacement with halide ions') and on the reverse you write the chemical equations that you must know.

Uses of chlorine and chlorate(I) compounds

Revised

Reaction of chlorine with water

Chlorine reacts with water according to the equation:

chlorine + water → chloric(I) acid + hydrochloric acid

$$Cl_2(g) + H_2O(l) \rightarrow HOCl(aq) + HCl(aq)$$

- The chloric(I) acid formed, $HOCl(aq)$, is a powerful oxidising agent. Chlorine is in oxidation state +1.
- In the above reaction, the chlorine initially is in oxidation state 0. At the end of the reaction, chlorine is in oxidation state +1 in chloric(I) acid (an oxidation); in hydrochloric acid it is in oxidation state −1 (a reduction).
- The powerful oxidising nature of chloric(I) acid is exploited in using chlorine to **purify water supplies** for drinking.
- Although chlorine is poisonous, its benefits in killing microorganisms responsible for water-borne diseases such as cholera, typhoid and diphtheria outweigh any potential disadvantages.
- Chloric(I) acid, $HOCl(aq)$, is a bleaching and oxidising agent, and it decomposes in sunlight to form oxygen:

  $2HOCl(aq) \rightarrow 2HCl(aq) + O_2(g)$

Chlorine has been simultaneously oxidised and reduced in a process known as **disproportionation**.

Typical mistake

Many candidates know that chlorine is added to water to disinfect it; but few understand that its action is due to the oxidising nature of the chloric(I) acid formed.

Reaction of chlorine with cold, aqueous, dilute sodium hydroxide solution

Chlorine disproportionates in cold, dilute and aqueous alkali:

$Cl_2(g) + 2NaOH(aq) \rightarrow NaOCl(aq) + NaCl(aq) + H_2O(l)$

- Chlorine's oxidation state increases from 0 to +1 and also decreases from 0 to −1. The product of this reaction, sodium chlorate(I) (NaOCl), is used as a **bleach** and as a **disinfectant** to kill germs.
- The chlorate(I) ion **oxidises** germs and bacteria by removing electrons:

  $ClO^-(aq) + 2H^+(aq) + 2e^- \rightarrow Cl^-(aq) + H_2O(l)$

Now test yourself

Tested ☐

5 Describe a chemical test to distinguish between lithium chloride and lithium iodide. Write ionic equations for the reactions that take place.

6 **(a)** Give different uses for chlorine and sodium chlorate(I).

(b) Give reagents for a reaction in which chloric(I) acid forms.

Answers on p. 105

Group 2 elements, the alkaline Earth metals

The elements in group 2 — beryllium, magnesium, calcium, strontium, barium and radium — are all metals.

Trends in physical properties

Revised

Atomic radius

The atomic radius **increases** down the group as, shown in Figure 10.2.

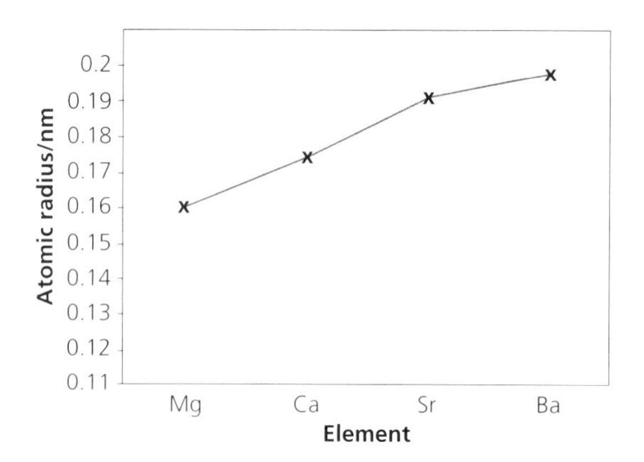

Figure 10.2 Atomic radii of group 2 elements

- The number of electron energy levels increases down the group.
- The outer electrons experience more shielding.
- There will be a weaker electrostatic attraction between the outer electrons and the nucleus.
- So the atomic radius will increase.

Melting points of the elements

As Figure 10.3 shows, the general trend in melting point is for this to **decrease** down the group.

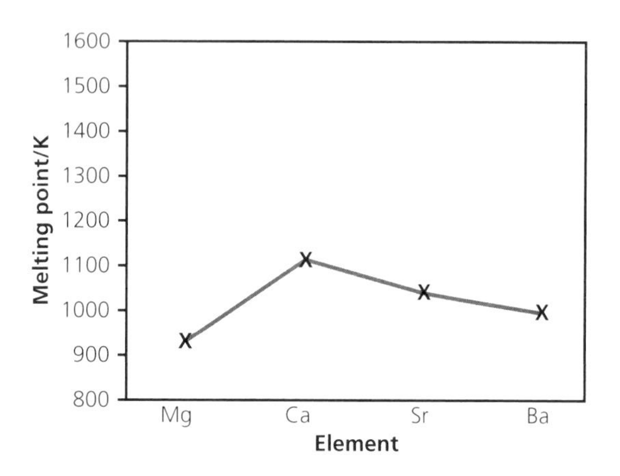

Figure 10.3 Melting points of group 2 elements

Magnesium's melting point does not fit the general trend because it has a different metallic structure from the other elements in the group.

- All the elements are metals so the structure of all these elements is a **lattice metallic**.
- Each atom donates two electrons to the delocalised electrons in the metallic structure.
- Ions of charge +2 form in the structure and these are bonded electrostatically by the mobile electrons.
- As the group is descended, the radius of the +2 ion increases because there are more electron energy levels.

- The outer electrons are further from the nucleus of each ion, and therefore further from the delocalised mobile electrons.
- The strength of the metallic bonding therefore decreases, and the melting point falls.

The **first ionisation energy** is the energy required to remove 1 mole of electrons from 1 mole of gaseous atoms under standard conditions:
$M(g) \rightarrow M^+(g) + e^-$

First ionisation energy of the elements

As Figure 10.4 shows, the first ionisation energy **decreases** down the group:

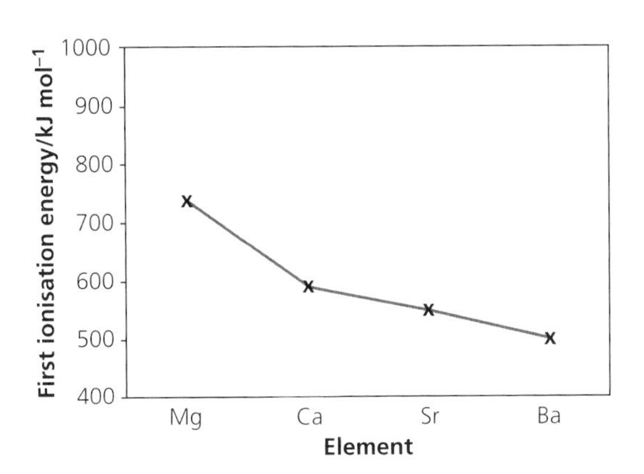

Figure 10.4 First ionisation energy of group 2 elements

- There are more electron energy levels as the group is descended.
- The outer electron being removed is progressively further from the nucleus.
- There will be less electrostatic attraction between the outer electron being removed and the nucleus.
- Therefore the first ionisation energy will decrease.

Reactions of group 2 elements and their compounds

Revised

All the metals are reactive, but not as reactive as group 1 metals. They all react by losing their outer two s-electrons to form a positive ion (a cation) with a +2 charge.

Reaction with water

Group 2 metals react with cold water according to the general equation (where M is the group 2 metal):

$$M(s) + 2H_2O(l) \rightarrow M(OH)_2(aq) + H_2(g)$$

A solution of the metal hydroxide (a weak base) forms along with hydrogen gas.

Reactivity **increases** down the group as it becomes progressively easier to remove two electrons because both the distance and the shielding between the outer electrons and the nucleus increase.

So the reactivity order for the metals, based on their rate of reaction with water is:

$$Mg \ll Ca < Sr < Ba$$

Example

Write equations for the reactions of barium and magnesium with cold water. Comment on any differences in the observed rates of reaction.

Answer

- Barium: $Ba(s) + 2H_2O(l) \rightarrow Ba(OH)_2(aq) + H_2(g)$
- Magnesium: $Mg(s) + 2H_2O(l) \rightarrow Mg(OH)_2(aq) + H_2(g)$

The reaction of magnesium with water is much slower than that of barium. In fact, magnesium reacts only very slowly with cold water.

Group 2 metals also react with steam, in which the reaction rate is considerably faster than that with cold water. Magnesium reacts violently with steam to form magnesium **oxide** as shown below:

$$Mg(s) + H_2O(g) \rightarrow MgO(s) + H_2(g)$$

- The metal hydroxide that forms in the reaction with cold water becomes more soluble as the group is descended.
- Because the hydroxide is more soluble as the group is descended, the concentration of hydroxide ions in solution will be higher. The alkalis that are formed are stronger.
- The order of the relative solubility of the metal hydroxides and their strength as soluble bases (alkalis) is:

  $$Mg(OH)_2 < Ca(OH)_2 < Sr(OH)_2 < Ba(OH)_2$$
- Magnesium hydroxide is used as an **antacid in medicine** — it reacts with excess acid in the stomach and relieves indigestion. Calcium hydroxide is used in agriculture because it **neutralises excess acidity in soil**. Both uses exploit the compound's ability to react as a base.

Group 2 sulfates

Down the group, the sulfates become less soluble. The order of solubility is:

$$MgSO_4 > CaSO_4 > SrSO_4 > BaSO_4$$

When magnesium sulfate is added to water, a colourless solution is formed. When barium sulfate is added to water, almost none dissolves because it is extremely insoluble.

Hydrochloric acid is added to decompose any carbonate that may be present. A carbonate will form a white precipitate with barium chloride solution and confuse the observations being made.

- Barium sulfate, $BaSO_4$, is an insoluble compound and its formation as a precipitate is easy to observe when **testing for a sulfate** because it forms a thick white precipitate.
- If sodium sulfate is the substance being tested, the reaction will be:

  $$Na_2SO_4(aq) + BaCl_2(aq) \rightarrow BaSO_4(s) + 2NaCl(aq)$$
- The **ionic equation** will be:

  $$Ba^{2+}(aq) + SO_4^{2-}(aq) \rightarrow BaSO_4(s)$$
- Barium sulfate is used in **medicine** when examining a patient's digestive system. The patient drinks a chalky barium sulfate suspension. When the patient is X-rayed, the barium sulfate coating inside the digestive tract absorbs a large proportion of the radiation. This highlights the black-and-white contrast of the X-ray photograph, so that doctors can diagnose digestive problems better.

The test for sulfate ions, SO_4^{2-} is to add hydrochloric acid and barium chloride solution. A white precipitate of barium sulfate will result.

Now test yourself

Tested

7 Write an equation to show how calcium reacts with water. What would be observed in this reaction?
8 Describe a chemical test that would enable you to distinguish between sodium nitrate and sodium sulfate.

Answers on p. 105

Exam practice

1 Concentrated sulfuric acid is added separately to potassium chloride and potassium iodide.
(a) **(i)** Write an equation for the reaction of concentrated sulfuric acid with potassium chloride.
(ii) Write an equation for the reaction of concentrated sulfuric acid with potassium iodide in which sulfur is a product.
(b) Predict two expected observations when concentrated sulfuric acid is added to potassium astatide (KAt). [2]
(c) Explain these trends down group 7 using the electronic structures of the atoms to help you:
(i) electronegativity [2]
(ii) boiling point [2]
(iii) atomic radius [2]
(d) Write equations for these reactions involving halogens or halide ions:
(i) potassium with fluorine gas [1]
(ii) sodium bromide solution with chlorine gas [1]
(iii) chlorine gas with water [1]
(iv) sodium chloride solution with silver nitrate solution [1]

2 **(a)** Give one use each for the following substances:
(i) calcium hydroxide, $Ca(OH)_2$ [1]
(ii) magnesium hydroxide, $Mg(OH)_2$ [1]
(iii) chlorine, Cl_2 [1]
(iv) barium sulfate, $BaSO_4$ [1]
(b) Describe chemical tests to distinguish between each of the compounds in these pairs:
(i) potassium iodide and potassium bromide [3]
(ii) copper(II) chloride and copper(II) sulfate [3]

Answers and quick quiz 5 online

Online

Examiners' summary

You should now have an understanding of:

- how electronegativity, melting point and boiling point vary down group 7 of the periodic table
- the oxidising power of the halogens
- halogen–halide displacement reactions
- the relative reducing power of the halides shown in their reactions with halogens and with concentrated sulfuric acid
- how to test for a halide ion using acidified silver nitrate solution
- how chlorine reacts with water and cold, dilute, aqueous sodium hydroxide
- how atomic radius, melting point and first ionisation energy vary down group 2 of the periodic table
- reactions of group 2 metals with water
- the variation in the solubility of group 2 hydroxides
- how to test for sulfate ions using acidified barium chloride solution
- the variation in solubility of group 2 sulfates

11 Extraction of metals

Principles of metal extraction

Converting sulfide ores to oxides

Revised

Metals are highly important to the world in which we live. They are found in the Earth in ores, usually as **oxides** or **sulfides**.

Sulfides are converted to oxides by roasting in air. For example, zinc sulfide, ZnS, reacts as follows:

$$2ZnS(s) + 3O_2(g) \rightarrow 2ZnO(s) + 2SO_2(g)$$

The problem with roasting is that sulfur dioxide, SO_2, is formed. If released into the atmosphere it could form an acid in lakes and could attack buildings made of a basic material like cement or limestone.

The sulfur dioxide formed by roasting sulfide ores is used to make sulfuric acid. This is done by reacting it with water and oxygen:

$$2SO_2(g) + 2H_2O(l) + O_2(g) \rightarrow 2H_2SO_4(aq)$$

Taking a metal oxide and extracting the metal from it involves the conversion of metal ions into atoms. This process involves the gain of electrons, and is therefore **reduction**. For example, extracting zinc from zinc oxide, ZnO, requires this reduction process:

$$Zn^{2+} + 2e^- \rightarrow Zn$$

This reduction can be carried out using **reducing agents** like **carbon** or **carbon monoxide**. These are both effective as reducing agents, as well as being relatively cheap.

Iron, **manganese** and **copper** are metals that can be formed by reducing their oxides with carbon or carbon monoxide. **Iron** is extracted by reduction of iron(III) oxide with carbon or carbon monoxide in a blast furnace:

$$Fe_2O_3(s) + 3C(s) \rightarrow 2Fe(l) + 3CO(g)$$

and/or:

$$Fe_2O_3(s) + 3CO(g) \rightarrow 2Fe(s) + 3CO_2(g)$$

Manganese is formed by reducing manganese(IV) oxide with carbon:

$$MnO_2(s) + C(s) \rightarrow Mn(s) + CO_2(g)$$

Copper is formed by heating malachite, $CuCO_3$. The copper(II) oxide formed is then reduced with carbon:

$$CuCO_3(s) \rightarrow CuO(s) + CO_2(g)$$

Then:

$$2CuO(s) + C(s) \rightarrow 2Cu(s) + CO_2(g)$$

Examiners' tip

Remember that when sulfides are roasted with air, both the metal and the sulfur form oxides.

Some metals like titanium, aluminium and tungsten cannot be formed by carbon reduction of their oxides because the metals are more reactive than carbon. Also, metal compounds with carbon can be formed instead. Other methods have to be found to extract these metals.

Aluminium extraction

Revised

Aluminium is extracted from an ore called bauxite, which is impure aluminium oxide, Al_2O_3. The main impurities in bauxite are silicon dioxide and iron(III) oxide.

Before aluminium oxide is electrolysed, it must be separated from the impurities in the bauxite ore. The purified aluminium oxide is dissolved in molten cryolite (Na_3AlF_6) at 950°C and is then ready for electrolysis. Dissolving the aluminium oxide in molten cryolite produces free ions at a temperature much lower than the melting point of pure aluminium oxide (2045°C). The use of cryolite also reduces energy costs.

Typical mistake

Many candidates think that cryolite is a catalyst for the process. It is not — it lowers energy costs by forming a mixture with aluminium oxide with a lower melting point by acting as a **solvent** for aluminium oxide to free the ions at a temperature much lower than the melting point pure aluminium oxide.

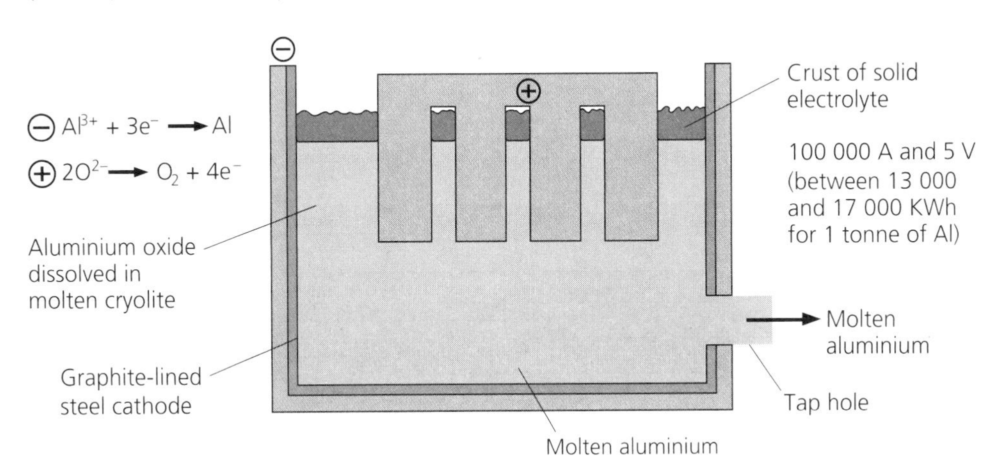

Figure 11.1 The Hall–Heroult cell

During the electrolysis (Figure 11.1) the positively charged and negatively charged ions in the aluminium oxide are free to move in the molten cryolite and migrate towards the oppositely charged electrode, where they are discharged.

- The **cathode** is made of steel and it is lined with graphite. Molten aluminium forms at the cathode by reduction of aluminium ions:

  At cathode: $Al^{3+}(l) + 3e^- \rightarrow Al(l)$
- The **anodes** are made of graphite. Oxygen is formed at the anode by oxidation of oxide ions:

  At anode: $2O^{2-}(l) \rightarrow 2O_2(g) + 4e^-$
- The carbon anodes need replacing regularly (this increases the cost) because they are eaten away due to reaction with the oxygen formed: $C(s) + O_2(g) \rightarrow CO_2(g)$ and $2C(s) + O_2(g) \rightarrow 2CO(g)$

The extraction of aluminium is only economic where cheap electricity is available, normally from hydroelectric power or nuclear power stations.

Revision activity

Revision flash cards are very useful in this section of work — write the name of the metal to be extracted on one side, and on the reverse write the key features: equations, conditions, advantages/disadvantages for the particular method of extraction.

Titanium extraction

Revised

The extraction of titanium from an ore containing the metal consists of two main stages.

1 Formation of the covalent chloride of titanium

This is achieved by mixing the ore of titanium(IV) oxide with **carbon** and heating the mixture in a stream of **chlorine gas**:

$TiO_2(s) + C(s) + 2Cl_2(g) \rightarrow TiCl_4(l) + CO_2(g)$

The volatile titanium(IV) chloride is then extracted from the reaction mixture using fractional distillation.

2 Production of the metal from titanium(IV) chloride

Either magnesium or sodium is used to displace titanium from the chloride. This reaction is carried out in a steel vessel at 700°C under an argon atmosphere:

$TiCl_4(l) + 2Mg(s) \rightarrow Ti(s) + 2MgCl_2(s)$

The process is expensive because of the high temperatures required and also due to the use of sodium or magnesium and argon. It is carried out as a **batch process** rather than a continuous process to make it when required.

Tungsten extraction

Revised

Tungsten cannot be extracted using carbon because tungsten carbide (WC) is formed rather than tungsten.

Tungsten(VI) oxide (WO_3) is heated with hydrogen at 900°C:

$WO_3(s) + 3H_2(g) \rightarrow W(s) + 3H_2O(l)$

There are obvious risks when using hydrogen because it is highly flammable in air, and potentially explosive.

Environmental aspects of metal extraction

Recycling metals

Revised

Recycling metals is important because the reserves of metal ores in the Earth's crust are decreasing, and energy costs are rising.

Recycling metals has many advantages:

- it creates less waste than conventional mining
- it saves resources like metal ores
- it saves energy resources compared to normal extraction techniques — for example, recycling aluminium from waste objects uses only 5% of the energy required to extract the same amount using electrolysis
- air pollution is reduced because gases like SO_2 (acid rain), CO (toxic), CO_2 (global warming threat) are not produced.

One **disadvantage** of the recycling process is the cost of collecting, transporting and sorting waste metals.

Recycling copper

Revised

Recycling copper from waste products is becoming very important because high grade copper ores are becoming scarce, and copper requires considerable energy to extract using carbon reduction of the oxide:

$$CuO(s) + C(s) \rightarrow CO_2(g) + Cu(s)$$

Carbon dioxide is a pollutant, the required temperatures are high and mining costs are very expensive.

Using scrap iron

An alternative method of obtaining copper is to use scrap iron — another waste product that is relatively cheap. The method is cheaper than extracting copper from its ore using vast amounts of energy. The scrap iron can be used to form copper in a redox reaction.

Scrap iron is mixed with a solution of copper(II) ions to form copper — the following redox reaction takes place:

$$CuSO_4(aq) + Fe(s) \rightarrow Cu(s) + FeSO_4(aq)$$

Ionic equation:

$$Cu^{2+}(aq) + Fe(s) \rightarrow Cu(s) + Fe^{2+}(aq)$$

This is a good method for extracting copper from low grade ores — the copper(II) solution can be concentrated prior to adding the iron.

Now test yourself

1 **(a)** Write half-equations for the reactions taking place on the surface of (i) the anode and (ii) the cathode during the electrolysis of aluminium oxide.
(b) Why is aluminium oxide added to cryolite before electrolysis?

2 Give three advantages of recycling to obtain metals, rather than using other techniques such as mining.

Answers on p. 105

Exam practice

1 Hydrogen is used to form tungsten industrially by reacting it with tungsten(VI) oxide, WO_3

(a) State the role of hydrogen in this reaction. [1]

(b) Write an equation for the reaction. [1]

(c) State one risk of using hydrogen gas in metal extractions. [1]

2 Scrap iron can be used to extract copper from dilute aqueous solutions containing copper(II) ions.

(a) Explain why this is a low-cost method of extracting copper. [1]

(b) Write the simplest ionic equation for the reaction of iron with copper(II) ions in aqueous solution. [1]

Answers and quick quiz 5 online

Online

Examiners' summary

You should now have an understanding of:

- how oxides are prepared by roasting prior to the reduction stage of extraction
- how potential pollutants like sulfur dioxide can be usefully employed for making other chemicals
- how the reduction of iron, manganese and copper ones involves carbon or carbon monoxide as reducing agents
- how aluminium is extracted from purified aluminium oxide using electrolysis
- how titanium is extracted from titanium(IV) oxide by a batch process involving reaction with carbon and chlorine, followed by reduction with either sodium or magnesium
- how tungsten is formed by hydrogen reduction of tungsten(VI) oxide
- how copper can be extracted from aqueous solutions containing copper (II) ions using scrap iron, and the environmental advantages of such a process

12 Haloalkanes

Haloalkanes can be classed as **primary**, **secondary** or **tertiary** depending on how many alkyl groups (methyl-, ethyl-, propyl-) are attached to the carbon atom that is bonded to the halogen atom. The haloalkanes shown in Figure 12.1 are classed as primary, secondary and tertiary respectively.

Haloalkanes are organic molecules based on alkanes in which one or more of the hydrogen atoms have been replaced by halogen atoms.

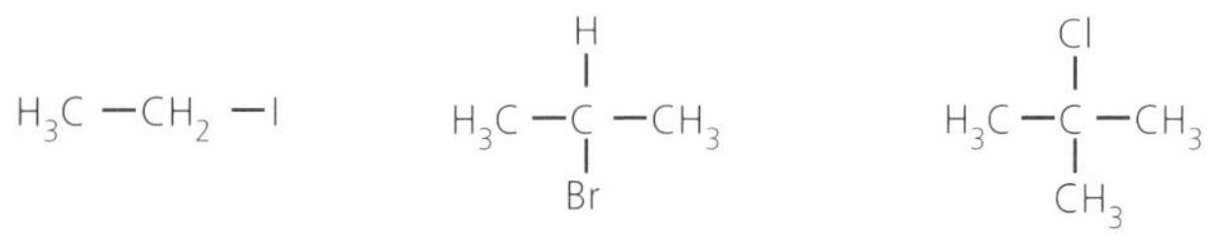

H_3C-CH_2-I — iodoethane

$H_3C-CH(Br)-CH_3$ (H above C, Br below C) — 2-bromopropane

$H_3C-C(Cl)(CH_3)-CH_3$ (Cl above C, CH_3 below C) — 2-chloro-2-methylpropane

Figure 12.1 Examples of haloalkanes

Synthesis of chloroalkanes

Chloroalkanes are compounds in which some hydrogen atoms of alkanes have been replaced by one or more chlorine atoms.

Methane with chlorine

Revised

Methane, CH_4, reacts with chlorine, Cl_2, in the presence of ultraviolet light in a reaction **mechanism** called **free-radical substitution**.

The overall equation for the reaction is:

$$CH_4(g) + Cl_2(g) \rightarrow CH_3Cl(g) + HCl(g)$$

The mechanism occurs in three stages.

- **Initiation:** Chlorine **radicals** are formed in the presence of ultraviolet light:

  $Cl_2 \rightarrow 2Cl\bullet$
- **Propagation:** chlorine radicals react with methane molecules to form new radicals and molecules. Adding the propagation steps together gives the overall equation for the reaction:

  $CH_4 + Cl\bullet \rightarrow HCl + \bullet CH_3$

then:

$\bullet CH_3 + Cl_2 \rightarrow CH_3Cl + Cl\bullet$

- **Termination:** radicals react with each other to form molecules:

  $\bullet CH_3 + Cl\bullet \rightarrow CH_3Cl$ or $\bullet CH_3 + \bullet CH_3 \rightarrow C_2H_6$

Radicals are molecules, atoms or ions that have an unpaired electron. This makes them very reactive.

Further substitutions, in which more hydrogen atoms are replaced by halogens, are possible. Using excess alkane will limit the extent of this happening.

Haloalkanes, such as chloroalkanes and chlorofluoroalkanes, can be used as solvents.

Ozone depletion

Revised

Ozone gas, O_3, is formed in the upper atmosphere of our planet. It protects life on Earth from potentially harmful ultraviolet radiation from the Sun.

Chlorine radicals, Cl•, react with ozone resulting in its removal or depletion. Chlorine radicals are formed in the upper atmosphere from haloalkanes containing C–Cl bonds, for example the chlorofluorocarbon (CFC) dichlorodifluoromethane, CCl_2F_2

The reaction involves a sequence of initiation and propagation steps.

- **Initiation:** formation of chlorine radicals from the CFC in the presence of ultraviolet light:

  $CCl_2F_2 \rightarrow •CClF_2 + Cl•$
- **Propagation:** reaction of chlorine radicals with ozone molecules

  $Cl• + O_3 \rightarrow ClO• + O_2$

then:

$ClO• + O_3 \rightarrow 2O_2 + Cl•$

Adding these two propagation equations gives the overall equation for the reaction:

$2O_3(g) \rightarrow 3O_2(g)$

Chlorine radicals catalyse the decomposition of ozone. Notice how they are used up in the reaction and then reformed at the end, this being typical catalytic behaviour. This means that chlorine radicals can continue to destroy ozone for many years.

Hydrofluoroalkanes, HFCs, and hydrofluorohydrocarbons, HCFCs, are now being used as alternatives to CFCs. However, HCFCs can still deplete the ozone layer. Although their depleting effect is only about one-tenth that of CFCs, serious damage is still being caused. HCFCs are a short-term fix until better replacements can be developed.

Now test yourself

Tested

1 **(a)** Explain why CFCs are damaging to the ozone layer.
  (b) State what is meant by a 'radical'.
  (c) Give two equations to show how chlorine radicals attack ozone.

2 Ethane gas reacts with bromine in the presence of ultraviolet light. Bromoethane is one of the products of the reaction.
  (a) Write an overall equation for the main reaction taking place.
  (b) Name the mechanism by which the reaction takes place.
  (c) Outline a likely mechanism for the reaction.

Answers on p. 106

Nucleophilic substitution

Polar molecules

Revised

Haloalkanes are **polar molecules**. They contain electronegative halogen atoms, X, that are attached to carbon atoms and the bond is polarised:

$^{\delta+}C-X^{\delta-}$

Haloalkanes may react by having their carbon–halogen bond broken. The ease of this breaking depends on the **strength** of the covalent bond.

In the case of **nucleophilic substitution**, a **nucleophiles** attacks the slightly positively charged carbon atom. The halogen atom then is substituted by the nucleophile.

A **nucleophile** is a donor of a lone pair of electrons.

Common nucleophiles include hydroxide, **:**OH^-; cyanide, **:**CN^- and ammonia, **:**NH_3

The ease of carbon–halogen bond breaking is:

C–I > C–Br > C–Cl ≫ C–F

Therefore iodoalkanes react faster than other haloalkanes because the C–I bond is the weakest.

The order of reactivity is:

iodoalkanes > bromoalkanes > chloroalkanes ≫ fluoroalkanes

Nucleophilic substitution reactions are highly important in organic synthesis because replacing a halogen atom with another functional group is useful and convenient.

Typical mistake

It is a common misconception for candidates to explain the rate of reaction of haloalkanes with nucleophiles in terms of C–X bond polarity, when it is in fact **bond strength** that is key.

Reaction with aqueous sodium hydroxide solution

Revised

Haloalkanes undergo substitution reactions with hot, aqueous sodium hydroxide solution in which the $:OH^-$ ion from the alkali acts as a **nucleophile** (lone-pair donor) and attacks the carbon atom attached to the halogen, which is lost as a halide ion (see Figure 12.2). For example, using chloromethane:

$$CH_3Cl + OH^- \rightarrow CH_3OH + Cl^-$$

methanol

H
H—C—Cl (δ+, δ−)
H
HO:
→
H—C—OH + Cl⁻

Figure 12.2 The nucleophilic substitution of chloromethane by hydroxide ions

In this reaction, the hydroxide ion nucleophile donates a lone pair of electrons to the carbon atom with the δ+ charge in the haloalkane. This is shown using a 'curly arrow' in the mechanism. The two bonding electrons in the carbon–halogen bond then move onto the halogen atom, and a halide ion forms, again shown by a curly arrow.

Reaction with sodium cyanide

Cyanide ions, :CN⁻, from sodium cyanide act as nucleophiles towards haloalkanes such as iodomethane:

$$CH_3I + CN^- \rightarrow CH_3CN + I^-$$

ethanenitrite

The mechanism for this process using the cyanide nucteophile (Figure 12.3) is identical to the previous case using hydroxide ions, OH^-

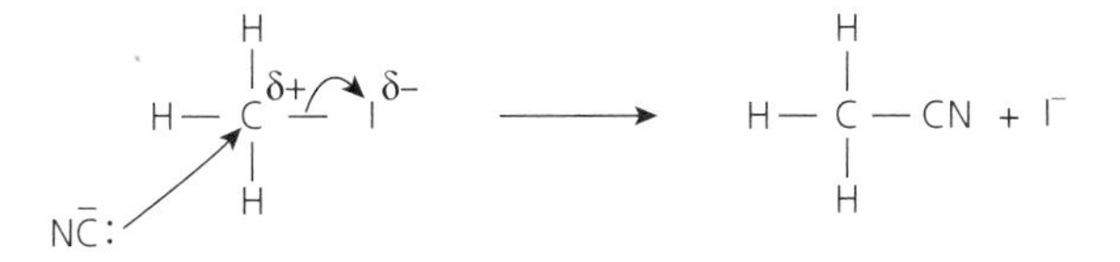

Figure 12.3 Cyanide ions reacting with iodomethane

The reaction with ethanolic ammonia

An ammonia molecule acts as a nucleophile and donates its lone pair of electrons to the carbon atom attached to the halogen (Figure 12.4). For example:

$$C_2H_5Br + 2NH_3 \rightarrow C_2H_5NH_2 + NH_4Br$$

ethylamine

Figure 12.4 The nucleophilic reaction between ammonia and bromoethane

Hydrogen bromide is not formed in this reaction because a second ammonia molecule reacts as a base to remove a proton and form ammonium bromide.

This reaction requires **ethanol** as a solvent and needs an excess of ammonia to limit further substitution.

Elimination

Elimination reactions

When potassium hydroxide solution, KOH(aq), is added to 2-bromopropane, a nucleophilic substitution reaction in which the hydroxide ion acts as a nucleophile:

$$CH_3\text{–}CHBr\text{–}CH_3 + OH^- \rightarrow CH_3\text{–}CH(OH)\text{–}CH_3 + Br^-$$

propan-2-ol

However, it is also possible for the hydroxide ion to behave as a **base** — a proton is removed from one of the carbon atoms next to the carbon attached to the halogen atom. In this case, **an elimination reaction** takes place forming an alkene (Figure 12.5).

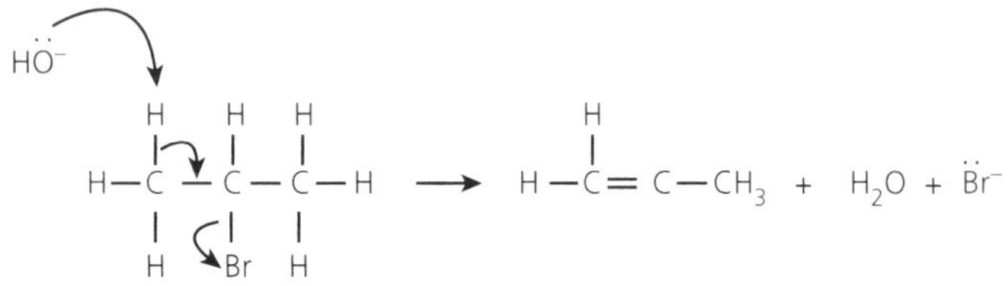

Figure 12.5 Mechanism for an elimination process

The solvent, ethanol, promotes the elimination process over the nucleophilic substitution process.

The products of this reaction are propene, water and bromide ions:

$$CH_3{-}CHBr{-}CH_3 + {}^{-}OH \rightarrow CH_3{-}CH{=}CH_2 + Br^{-} + H_2O$$

Examiners' tip

The choices of solvent and temperature have a crucial role in this type of reaction. A hot ethanol solvent promotes elimination; warm aqueous conditions promote the nucleophilic substitution.

Exam practice

1 **(a)** Name these haloalkanes. [4]

(i)

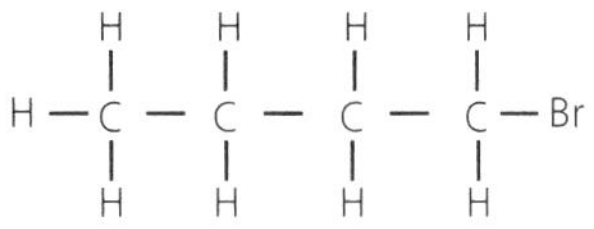

(ii)

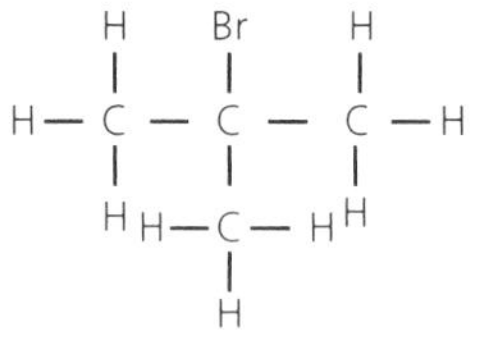

(iii)

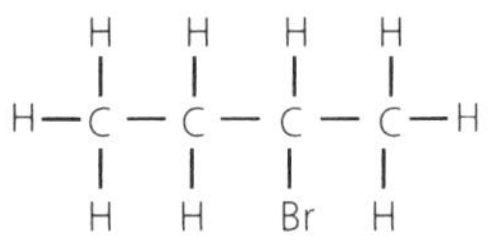

(iv)

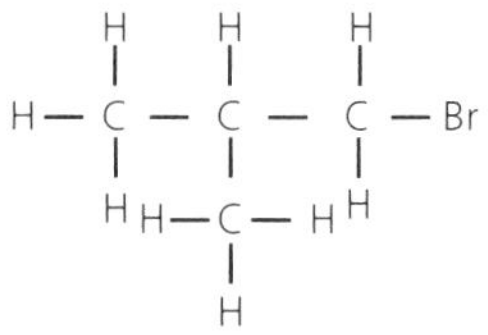

(b) Classify the haloalkanes in part (a) as primary, secondary or tertiary. [4]

(c) How are the four molecules related? [1]

(d) Draw structural formulae to show the products formed when 1-iodobutane reacts with:

(i) excess ammonia in ethanol [1]

(ii) warm NaOH(aq) [1]

(iii) NaCN(aq) [1]

(e) Draw the structure of the alternative product in part (d)(ii) if hot ethanol were used as the solvent [2]

(f) How would the rates of reaction differ when chloroethane and bromoethane are reacted separately with aqueous sodium hydroxide solution? Explain your answer. [2]

Answers and quick quiz 6 online

Online

Examiners' summary

You should now have an understanding of:

- how methane reacts with chlorine in the presence of ultraviolet light
- the free-radical substitution process
- how ozone is depleted by the presence of chlorine radicals
- nucleophilic substitution mechanisms
- how haloalkanes react with NaOH, KCN and excess NH_3
- how a hydroxide ion may behave either as a nucleophile or a base, and how the products differ depending on the conditions used
- elimination processes involving haloalkanes

13 Alkenes

Structure, bonding and reactivity

Carbon–carbon double bonds

Revised

Alkenes are hydrocarbons that contain carbon–carbon double bonds. For this reason they are described as being **unsaturated**.

The presence of a double bond makes alkenes **more reactive** than alkanes. This is because the double bond is an area of high electron density in the molecule.

The structure of the carbon–carbon double bond is shown in Figure 13.1.

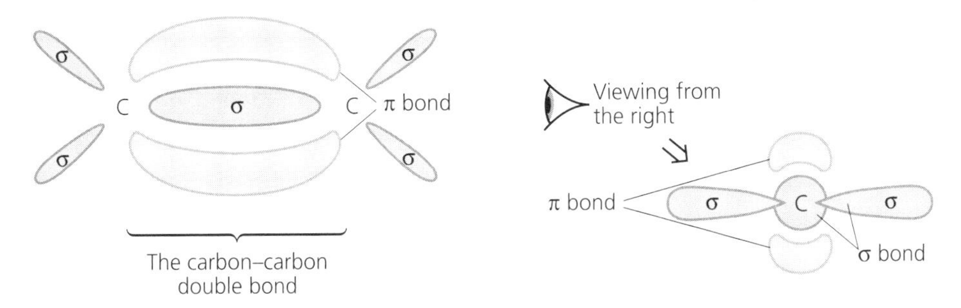

Figure 13.1 Representation of the σ and π bonds in a double bond

- The carbon–carbon double bond arrangement is **planar** (flat) and the internal bond angle is **120°**.
- The π bond **prevents rotation** about the σ bond.
- This lack of rotation gives rise to **E–Z (geometrical) isomers**.
- E–Z geometrical isomerism is one example of **stereoisomerism**.

> **Stereoisomers** are compounds with the same structural formula but with bonds arranged differently in space.

The E–Z isomers in Figure 13.2 show how molecules such as but-2-ene and 1,2-dichloroethene can exist in more than one form. A pair of E–Z isomers differ because of the different arrangement of groups in space.

E-but-2-ene: H, CH_3 / C=C / CH_3, H

Z-but-2-ene: H_3C, CH_3 / C=C / H, H

E-1,2-dichloroethene: H, Cl / C=C / Cl, H

Z-1,2-dichloroethene: Cl, Cl / C=C / H, H

Figure 13.2 Geometric isomers

The letter 'E' indicates that the groups are on opposite sides of the double bond, and 'Z' shows they are on the same side.

Now test yourself

Tested

1 The alkene pent-2-ene shows E–Z (geometrical) isomerism. Draw the displayed formulae of the two stereoisomeric forms.

Answers on p. 106

Addition reactions of alkenes

Electrophilic addition

Revised

Alkenes react with other substances in a process known as **addition**.

In these reactions, the electron pair present in the π bond of the double bond is donated to the attacking species to form a new single (sigma, σ) bond. The attacking species is therefore an electron–pair acceptor, or **electrophile**.

An **electrophile** is an acceptor of a lone pair of electrons.

The mechanism that describes reactions of this type is called **electrophilic addition**.

Bromine, Br_2, reacts with an alkene in this way. For example, ethene, C_2H_4, reacts with bromine to form 1,2-dibromoethane:

$H_2C{=}CH_2 + Br{-}Br \rightarrow H_2CBr{-}CH_2Br$

1,2-dibromoethane

The mechanism by which this reaction takes place is shown in Figure 13.3.

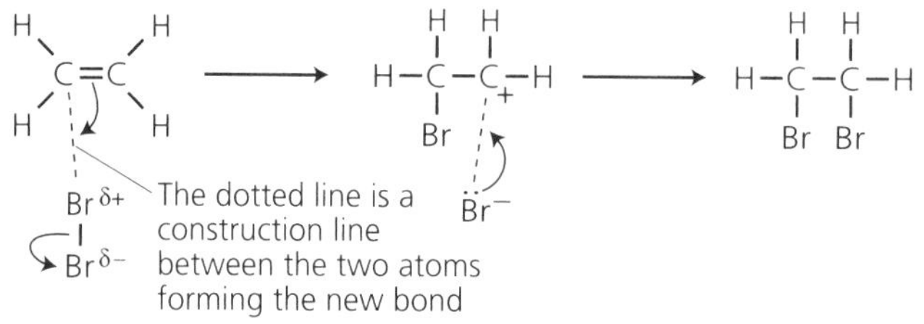

H H C=C H H → H H H−C−C−H Br + → H H H−C−C−H Br Br

Brδ+ Brδ− Br−

The dotted line is a construction line between the two atoms forming the new bond

Figure 13.3 An electrophilic addition reaction

The reaction of bromine with an alkene is used as a **chemical test for unsaturation**. In this test, the colour changes from **orange** to **colourless**.

Examiners' tip

It is one of the most important chemical tests to know — the use of bromine as a test for a carbon–carbon double bond. Know it well.

Reactions of alkenes

Revised

Alkenes undergo many reactions — for example the reaction with bromine just mentioned —the other key ones are as follows.

Hydrogen bromide

Hydrogen bromide adds to the double bond in the same way bromine does, via an electrophilic addition process.

For example, ethene, C_2H_4, reacts with hydrogen bromide to form bromoethane, C_2H_5Br

$H_2C{=}CH_2 + H{-}Br \rightarrow H_3C{-}CH_2Br$

In this mechanism (Figure 13.4), the first step involves protonation of the carbon–carbon double bond to form a **carbocation**. In the second step, the bromide ion attacks the carbocation.

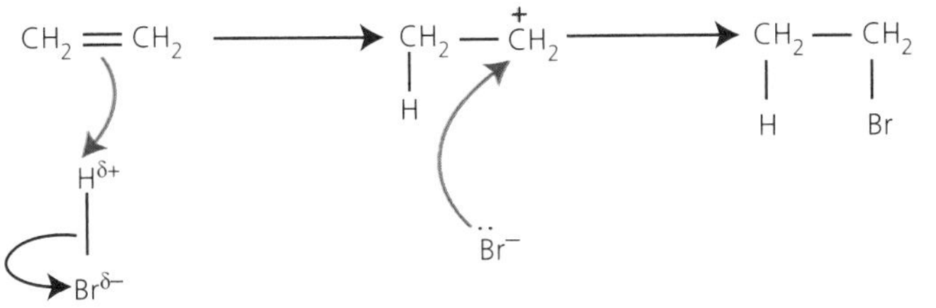

$CH_2{=}CH_2 \rightarrow CH_2(H){-}\overset{+}{C}H_2 \rightarrow CH_2(H){-}CH_2(Br)$

$H^{\delta+}{-}Br^{\delta-}$, $:Br^-$

Figure 13.4 Electrophilic addition of HBr to ethene

If hydrogen bromide is added to an unsymmetrical alkene, such as propene, two products can form: 1-bromopropane or 2-bromopropane (Figure 13.5). This is because either carbon atom can be protonated in the first step of the mechanism, leaving the bromide ion free to bond to either of the resulting **carbocations**.

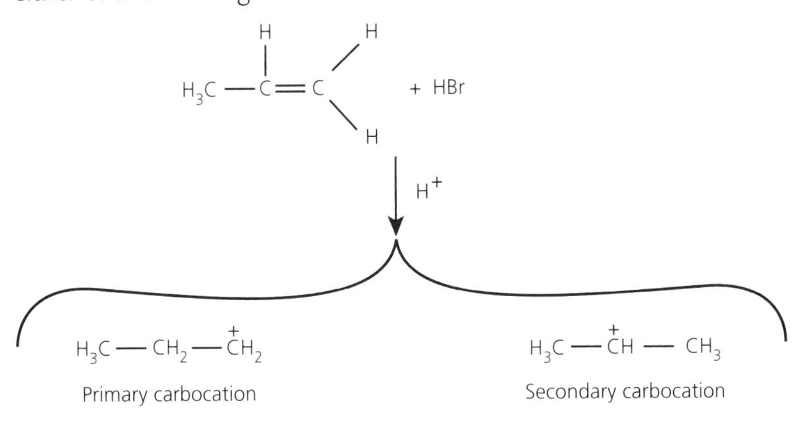

Figure 13.5 Electrophilic addition of HBr to propene

However, primary carbocations are less stable than secondary carbocations and so the latter are formed preferentially.

The bromide ion therefore reacts to form 2-bromopropane as the major product.

Example

Draw the product expected when but-1-ene reacts with hydrogen bromide.

Answer

- But-1-ene is $H_2C{=}CH{-}CH_2CH_3$
- The carbon atom coloured red below is attached to fewer hydrogen atoms than the other carbon atom in the double bond:

  $H_2C{=}CH{-}CH_2CH_3$

  In the product, this red carbon atom will be bonded to the bromine and the hydrogen will be bonded to the other carbon from the double bond.
- $H_3C{-}CHBr{-}CH_2CH_3$, 2-bromobutane, is the major product.
- This is formed via the more stable secondary carbocation:

$H_3C{-}\overset{+}{C}H{-}CH_2{-}CH_3$

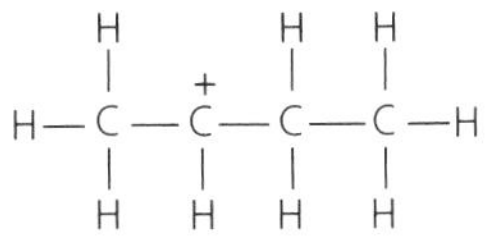

Now test yourself — Tested

(a) Draw the structure of 2-methylpropene.

(b) Draw the structures of the two carbocations that could form when 2-methylpropene reacts with hydrogen bromide.

(c) Which of the carbocations will be the more stable?

(d) Hence draw the structure of the major product expected in the reaction.

Sulfuric acid

Alkenes react with concentrated sulfuric acid at 0°C in an electrophilic addition process.

The products are hydrogensulfates:

$$C_2H_4 + H_2SO_4 \rightarrow \underset{\text{ethyl hydrogensulfate}}{H_3C\text{–}CH_2\text{–}OSO_3H}$$

The product is formed via a carbocation intermediate (Figure 13.6), the same as in the reaction with hydrogen bromide.

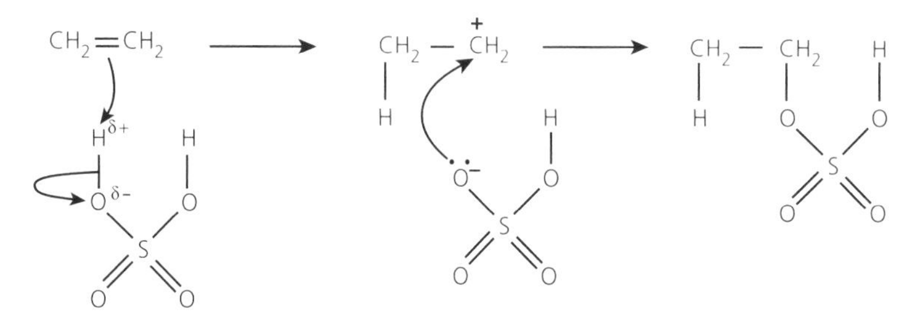

Figure 13.6 Electrophilic addition of H_2SO_4 to ethene

The product reacts with water — a process called hydrolysis — to form an **alcohol**. For example, ethyl hydrogensulfate forms ethanol:

$$H_3C\text{–}CH_2\text{–}OSO_3H + H_2O \rightarrow C_2H_5OH + H_2SO_4$$

Alcohols are produced **industrially** by the reaction of **steam** with an alkene in the presence of an acid catalyst. This process is called **hydration**.

Conditions used are 300°C and 60–70 atmospheres using phosphoric(v) acid as a catalyst. For example, ethanol can be made by the hydration of ethene using steam:

$$C_2H_4(g) + H_2O(g) \rightarrow C_2H_5OH(g)$$

Revision activity

There are several alkene reactions you need to know. Why not test your understanding by drawing the structure of any alkene, and then drawing the products that this molecule would form with HBr, Br_2, H_2SO_4 then water.

Polymerisation

Addition polymers

Revised

Alkenes will **polymerise** under high pressure and high temperature conditions using a range of catalysts. In this process the individual alkene molecules, called **monomers**, bond together (using their π bonds) to form an **addition polymer**.

For example, ethene, C_2H_4, forms poly(ethene) in which the two electrons in the π bonds of the monomer (ethene) form σ (single) bonds with other ethene units as shown in Figure 13.7.

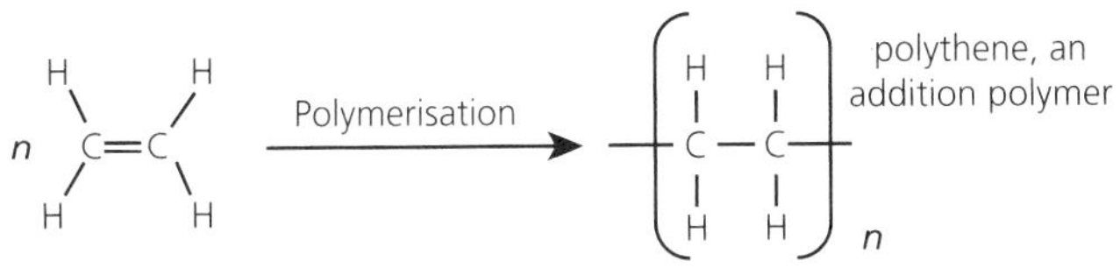

$$n\,CH_2{=}CH_2 \xrightarrow{\text{Polymerisation}} \left[\!-CH_2-CH_2-\!\right]_n$$

polythene, an addition polymer

Figure 13.7 Polymerisation

Examiners' tip

In most polymers, there are no carbon–carbon double bonds. Therefore, polymers are fairly unreactive substances.

Figure 13.8 shows how chloroethene (monomer) forms a polymer called poly(chloroethene) formerly known as polyvinylchloride (PVC). The polymer shown contains three repeating units.

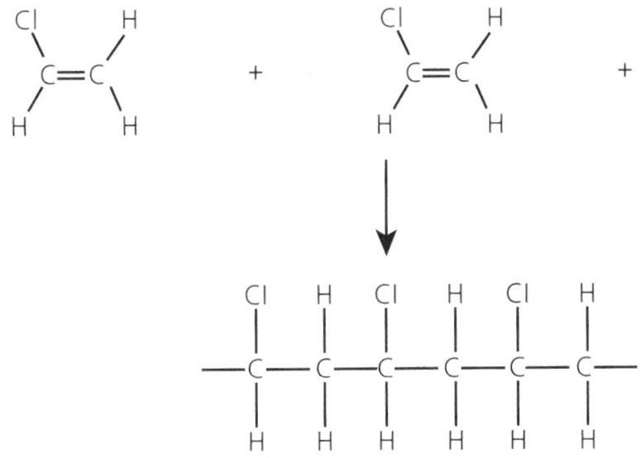

$$CHCl{=}CH_2 + CHCl{=}CH_2 + CHCl{=}CH_2 \rightarrow -CHCl-CH_2-CHCl-CH_2-CHCl-CH_2-$$

Figure 13.8 Making PVC

Polymers have many uses — film wrapping, carrier bags, bottles, kitchenware, shrink-wrap etc. However, disposal is a problem because many polymers do not biodegrade. Recycling poly(propene), for example, is a more cost-effective method than producing new polymer from crude oil. This is because energy costs are considerably less.

Now test yourself

Tested

3 Draw the structures of the major products expected when propene reacts with:

(a) hydrogen bromide, HBr

(b) bromine, Br_2

(c) sulfuric acid, H_2SO_4

(d) steam and an acid catalyst.

4 Draw the structure of the polymer poly(propene) showing three repeating units.

Answers on p. 106

Exam practice

1 **(a)** State what is observed when bromine water is added to **(i)** an alkene and **(ii)** an alkane. [2]

(b) Name the mechanism for the reaction between an alkene and a halogen such as bromine. [1]

(c) Draw the mechanism to show the reaction between propene and bromine. Take care when indicating the positions of any curly arrows in the mechanism. [4]

2 **(a)** Name these alkenes. [3]

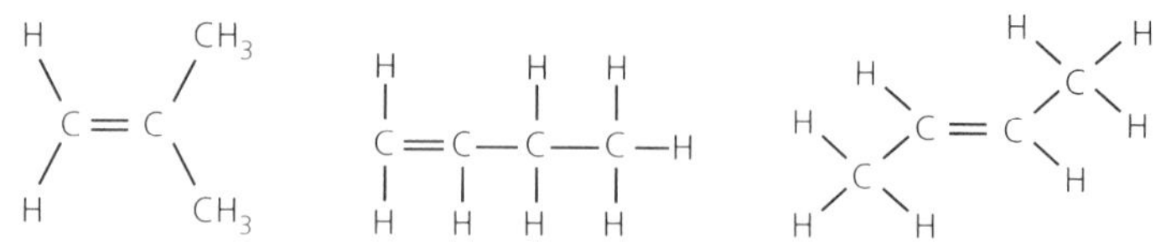

molecule 1 molecule 2 molecule 3

(b) Give the common molecular formulae for each of these alkenes. [1]

(c) How are all three molecules related? [1]

(d) Explain which of the molecules is able to exist as E–Z isomers. [2]

(e) Draw the structure of the major product formed when molecule 2 reacts with hydrogen bromide. Explain, using the mechanism of the reaction, why this is the major product. [3]

(f) Draw structures of the repeating units in the polymers formed by:

(i) molecule 2

(ii) molecule 3 [2]

Answers and quick quiz 6 online

Online

Examiners' summary

You should now have an understanding of:

- what is meant by the term 'unsaturated' when applied to alkenes
- the structure of a carbon–carbon double bond and how free rotation is restricted
- the existence of E–Z isomerism
- the reactions of alkenes with HBr, H_2SO_4 and Br_2
- how two products are possible when HBr reacts with an unsymmetrical alkene, and why one may be preferred over the other
- how an alcohol is produced industrially by the reaction of an alkene with steam in the presence of an acid catalyst
- how alkenes polymerise to form addition polymers

14 Alcohols

Naming alcohols

The number of carbon atoms

Revised

Alcohols are named according to the number of the carbon atoms in their molecules, and also the position of the hydroxyl group, OH.

Alcohols are hydrocarbons that have one or more of their hydrogen atoms replaced by a hydroxyl group (–OH).

The molecules in Figure 14.1 are called pentan-1-ol, pentan-2-ol and 2-methylbutan-2-ol respectively.

```
                                                              H
                                                              |
   H  H  H  H  H             H  H  H  OH H                H — C — H
   |  |  |  |  |             |  |  |  |  |             H  H   |   H
H—C—C—C—C—C—OH          H—C—C—C—C—C—H                  |  |   |   |
   |  |  |  |  |             |  |  |  |  |          H—C—C—C—C—H
   H  H  H  H  H             H  H  H  H  H             |  |   |   |
                                                       H  H   OH  H
```

Figure 14.1 Isomeric alcohols

Ethanol production

Ethanol, C_2H_5OH, is a very useful substance because it is a fuel that burns to release heat energy:

$$C_2H_5OH(l) + 3O_2(g) \rightarrow 2CO_2(g) + 3H_2O(l)$$

There are two very different ways of making ethanol: **fermentation of sugars** and **hydration of ethene**.

Fermentation of sugars

Revised

Sugars are obtained from crops such as sugar cane, sugar beet, corn, rice and maize. Naturally-occurring starches (in corn, rice and maize) form sugars by enzyme-controlled processes and, once the sugars are formed, fermentation takes place in the presence of **enzymes in yeast**:

$$C_6H_{12}O_6(aq) \rightarrow 2C_2H_5OH(aq) + 2CO_2(g)$$

Fermentation is the process in which **microorganisms** convert one substance into another, normally in the **absence of air**.

Ethanol formed in this way is called a **biofuel** because it is formed from **renewable** plants (those that can be grown again or replenished).

The advantages of this process are:

- sugars come from **renewable resources**
- the energy requirements are low (enzymes operate best at room or body temperature)
- the process is carbon-neutral — any carbon dioxide produced is removed from the air by photosynthesis in the original plant so, overall, there has been no net change in carbon dioxide levels.

- the process of fermentation is (usually) a batch process so the use of manpower is much greater than in the automated continuous hydration process.

Disadvantages include:

- the ethanol is in the form of an aqueous solution — therefore heat energy is needed for fractional distillation; producing this energy is expensive and not carbon-neutral
- the process is slow — it takes several days for the ethanol concentration to reach its limit of approximately 16% by volume
- renewable crops require huge surface areas of land to grow, and the machinery and people-power required in harvesting these crops can be significant.

Examiners' tip

Photosynthesis involves the removal of carbon dioxide from the air; fermentation produces carbon dioxide. So, carbon-neutral fuels have zero effect on carbon dioxide levels.

Hydration of ethene

Revised

This reaction has been discussed previously in Chapter 13 (see page 90).

Ethene can be hydrated in the presence of phosphoric(v) acid to form ethanol at 300°C and a pressure of 60–70 atmospheres.

$$\begin{array}{c} CH_2 \\ \| \\ CH_2 \end{array} + H_2O \xrightarrow[300^\circ C,\ 65\ \text{atm}]{H_3PO_4} \begin{array}{l} CH_3 \\ \| \\ CH_2\text{—}OH \end{array}$$

The advantages of this method are:

- the ethanol is pure, so no separation is required afterwards
- the rate of reaction is high, so the ethanol is formed quickly
- the process is continuous, so ethene and steam can be added automatically without any costly breaks in production or use of manpower.

Disadvantages include:

- ethene is obtained from the cracking of crude oil and cracking requires huge energy reserves
- crude oil is a non-renewable resource, so once the crude oil is used it can't be replenished or reformed.
- the process of hydration requires a high temperature and high pressure — both of which have considerable energy and cost implications.

Examiners' tip

Make sure that you know the disadvantages and advantages of the two methods of making ethanol, including the environmental issues.

Classification of alcohols

Alcohols are sorted into groups (Figure 14.2) according to how many alkyl groups are attached to the carbon atom bonded to the hydroxyl group.

```
     R                                     CH3
     |                                     |
H — C — OH      Primary alcohol       H — C — OH
     |           e.g. ethanol              |
     H                                     H

     R                                     CH3
     |                                     |
R'— C — OH      Secondary alcohol   H3C — C — OH
     |           e.g. propan-2-ol          |
     H                                     H

     R                                     CH3
     |           Tertiary alcohol          |
R'— C — OH   e.g. 2-methylpropan-2-ol H3C — C — OH
     |                                     |
     R"                                    CH3
```

Figure 14.2 Primary, secondary and tertiary alcohols

The class of alcohol determines the type of reaction it may undergo — particularly when being oxidised.

Reactions of alcohols

Revised

Alcohols can be oxidised using acidified potassium dichromate(VI) as the oxidising agent. The product molecule formed depends on the class of the original alcohol.

Primary alcohols

Primary alcohols are oxidised to form **aldehydes**, and then **carboxylic acids**. For example, ethanol is oxidised in the first stage to form **ethanal**, and then this is oxidised to form **ethanoic acid** — Figures 14.3 and 14.4 show the convenient use of '[O]' to represent the oxidising agent:

```
        H
        |                                 //O
CH3— C —OH  +  [O]  ——————>  CH3— C              +  H2O
        |                                 \
        H                                  H
    ethanal                        ethanal

          //O                                //O
CH3— C          +  [O]  ——————>  CH3— C
          \                                  \
           H                                  OH
    ethanal                       ethanoic acid
```

Figure 14.3 Oxidation of primary alcohols

During these reactions the orange dichromate(VI) ions are **reduced** and turn dark green. The alcohol is **oxidised**. These reactions are therefore **redox reactions**.

- Aldehydes are normally removed by **distillation** as the reaction proceeds (to prevent further oxidation).

- Carboxylic acids are formed in a process involving **reflux**.A secondary alcohol and acidified potassium dichromate(VI) are normally **refluxed** to ensure that the reaction goes to completion.

Secondary alcohols

As shown in Figure 14.4, secondary alcohols are oxidised to form **ketones**.

$$CH_3-CH(OH)-CH_3 + [O] \longrightarrow CH_3-C(=O)-CH_3 + H_2O$$

propan-2-ol → propanone

Figure 14.4 Oxidation of secondary alcohols

Tertiary alcohols

Tertiary alcohols cannot be oxidised using acidified potassium dichromate(VI) as an oxidising agent. So, no reaction occurs and the acidified potassium dichromate(VI) stays orange.

> **Examiners' tip**
> The symbol [O] represents the oxidising agent and is used to balance oxidation equations such as those shown. You should treat it as a normal element symbol, and make sure the oxygens balance.

1. Draw the displayed formula of pentan-1-ol.
2. Give the structures and names of the products formed when pentan-1-ol is oxidised using acidified potassium dichromate(VI).

Aldehydes and ketones

Aldehydes and ketones are called **carbonyl compounds** because they contain the carbonyl group. The functional groups present in aldehydes and ketones are shown in Figure 14.5 ('R' represents an alkyl group) and some specific molecules in Figure 14.6.

Figure 14.5 Functional groups present in aldehydes and ketones

Figure 14.6 Examples of aldehydes and ketones

Distinguishing aldehydes from ketones

Revised

Tollens' reagent

Aldehydes and ketones can be distinguished from one another using **Tollens' reagent**. Tollens' reagent is formed by mixing ammonia with silver nitrate solution.

In this reaction, ketones show no reaction, but aldehydes form a **silver mirror** on the inside of the test tube — the silver ions are **reduced**:

$$Ag^{+}(aq) + e^{-} \rightarrow Ag(s)$$

and the aldehyde is **oxidised** to a carboxylic acid. Using ethanal as an example:

$$CH_3CHO + [O] \rightarrow CH_3COOH$$

Fehling's solution

In this test, an aldehyde reduces copper(II) ions and an **orange-brown precipitate** of copper(I) oxide is formed. Ketones show no reaction.

Typical mistake

Many candidates focus on the reduction of the silver ion when using Tollens' reagent. However, don't forget that the aldehyde has been oxidised to a carboxylic acid in the reaction too.

Elimination

Dehydration reactions

Revised

Alcohols can be dehydrated using concentrated sulfuric acid (as a catalyst) to form an alkene. For example, propan-1-ol can be dehydrated to form propene:

$$CH_3CH_2CH_2OH \rightarrow CH_3CH{=}CH_2 + H_2O$$

Dehydration reactions can be used as a possible route for synthesising polymers from alcohols. For example, ethanol → ethene → poly(ethene).

So polymers could be made starting from sugars, avoiding the reliance on crude oil as the original source of carbon.

- Draw the structures of butanone and butanal.
- Describe a chemical test to distinguish butanone from butanal.

Answers on p. 106

Exam practice

1 A polymer such as poly(ethene) could be synthesised from glucose using this route:

$$\text{glucose: } C_6H_{12}O_6 \xrightarrow{\text{Step 1}} C_2H_5OH \xrightarrow{\text{Step 2}} C_2H_4 \xrightarrow{\text{Step 3}} \text{poly(ethene)}$$

(a) Give the reagents and conditions required for each stage of the route above. [6]

(b) Give the name of the chemical process in each step. [3]

(c) Draw the repeating unit in poly(ethene). [1]

2 The substance A below can be synthesised from an alcohol under appropriate conditions:

$CH_3{-}CH_2{-}CH(CH_3){-}CHO$ (drawn with CH_3 branch on the CH and C=O, C–H on the terminal carbon)

(a) Name molecule A. [1]

(b) Draw the structure of the alcohol that, when oxidised, would produce molecule A. [1]

(c) Give the name of the reagent(s) used to form A from the alcohol. [1]

Tollens' reagent is added to a sample of A and the resulting solution heated.

(d) (i) State what would be observed in the reaction. [1]

(ii) Draw the structure of the organic molecule formed in the reaction. [1]

Answers and quick quiz 6 online

Online

Examiners' summary

You should now have an understanding of:

- the molecular structure of alcohols and their key functional group
- how to name simple alcohols
- how ethanol can be made on an industrial scale, either by fermentation or hydration, and the advantages and disadvantages of each process
- how to classify alcohols as primary, secondary and tertiary
- the oxidation of alcohols of different classes using acidified potassium dichromate(VI)
- Tollens' reagent and how it is used to distinguish between aldehydes and ketones
- elimination reactions, in which alkenes are made from alcohols

15 Analytical techniques

Several modern analytical techniques are useful when determining the structures of molecules.

Mass spectrometry

Analysing organic compounds

Revised

Mass spectrometry instrumentation was discussed in Chapter 1 where it was used for analysing elements. It is also very useful when analysing organic compounds.

It is possible to measure with great precision the mass of a **molecular ion** if the relative isotopic masses to four decimal places are used, for example:

$^{1}H = 1.0078$ $^{12}C = 12.0000$ $^{14}N = 14.0031$ $^{16}O = 15.9949$

For example, two organic compounds, propane (C_3H_8) and ethanal (CH_3CHO), both have relative molecular mass $M_r = 44$ to the nearest whole number. Using a high-resolution mass spectrometer, two molecular ion peaks with the following more precise *m*/*z* values are obtained:

C_3H_8 44.0624 CH_3CHO 44.0261

When analysing samples, it is therefore possible to deduce the molecular formula of a compound by measuring its precise relative molecular mass. The relative molecular mass of a compound is given by the peak with the highest *m*/*z* value.

A computer database can have thousands of organic molecules' relative molecular masses measured to a high level of accuracy; a sample under test can then be compared to other spectra and its identity determined.

A **molecular ion** is the original molecule that has lost one electron, M^+. Its mass will be the virtually the same as the molecule being tested, because an electron has negligible mass.

Infrared spectroscopy

Atoms in covalent bonds are **vibrating** about a mean position. The frequency of the vibration depends on the masses of the atoms in the bond — the higher the mass of the atoms, the slower the vibrations.

Wavenumbers

Revised

The frequency of such a vibration can be quoted in **wavenumbers** — the number of waves in 1 cm. One particular bond will vibrate at a certain wavenumber. However, just as the energy of a particular covalent bond will vary slightly in different compounds, the frequency of vibration also varies, so values are usually quoted as a range as in Table 15.1.

Table 15.1 Typical wavenumbers in infrared spectra

Bond	Wavenumber/cm^{-1}
C–H in alkanes	2850–2960
C–C	750–1100
C=C	1620–1680
C–O	1000–1300
O–H in carboxylic acids	2500–3000
C=O	1680–1750
O–H in alcohols	3230–3550 (broad adsorption)
C–N in amines	1180–1360
C–Cl	600–800
C–Br	500–600

Figure 15.1 shows the displayed formula for ethyl ethanoate, $CH_3COOC_2H_5$.

$$CH_3 - \underset{\underset{\displaystyle O}{\|}}{C} - O - CH_2 - CH_3$$

Figure 15.1 Ethyl ethanoate

Notice the bonds present in this compound: C–H, C=O, C–O and C–C. These will vibrate at certain wavenumbers. For example, the C=O bond is expected to vibrate at 1743 cm^{-1} (Figure 15.2)

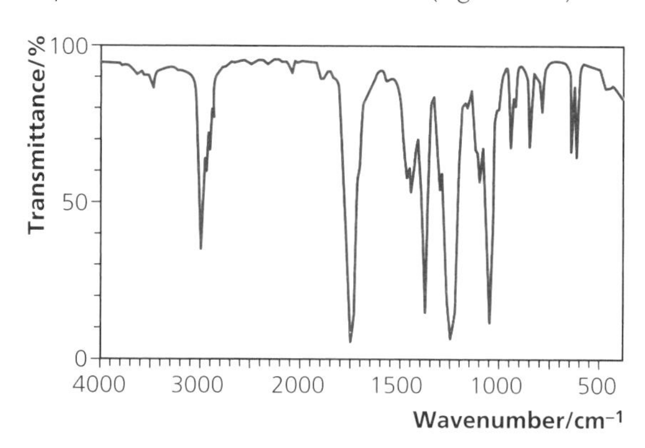

Figure 15.2 The infrared spectrum for ethyl ethanoate

Fingerprinting

The peaks below 1500 cm^{-1} in the spectrum are used collectively as a **fingerprint region**. This characteristic arrangement of peaks is unique to the compound, just as a fingerprint is unique to a person. Therefore by comparing the fingerprint region of an unknown compound with many on a computer database, the identity of the compound can be deduced.

It may be possible to detect impurities in a compound because they may give rise to absorptions that would not be expected for that compound.

Some molecules in the atmosphere absorb infrared radiation and contribute towards global warming. This is because the natural frequency of the bond vibration in molecules such as CO_2 or H_2O coincides exactly with the frequency of the infrared radiation. The same process takes place when a sample is put into an infrared spectrometer.

Tested

1. Draw the structure of propan-1-ol.
2. Write down the bonds present in propan-1-ol.
3. Give the wavenumbers of the main peaks expected in the infrared spectrum of this compound.
4. A small absorption is observed at a wavenumber of 1720 cm^{-1}. What does this suggest about the sample?

Answers on p. 107

Exam practice

1 In an experiment to prepare a sample of ethanal (CH_3CHO), ethanol (C_2H_5OH) is reacted with acidified potassium dichromate(VI) and the reaction mixture is distilled. Infrared spectra are obtained for ethanol and ethanal.

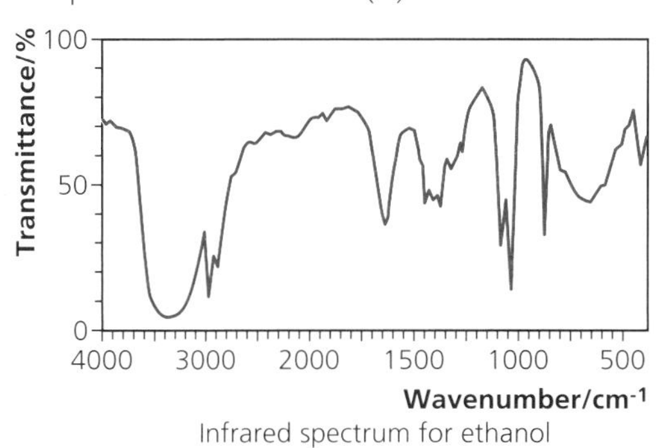

Infrared spectrum for ethanol

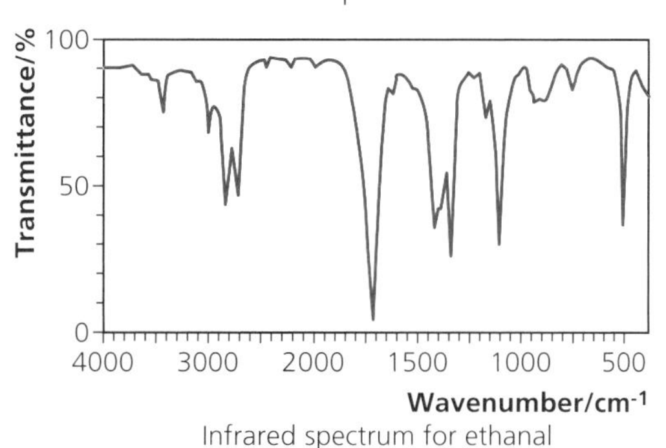

Infrared spectrum for ethanal

(a) Indicate the bonds that give rise to these absorptions:

(i) in the ethanol spectrum at 3400 cm^{-1}

(ii) in the ethanal spectrum at 1720 cm^{-1} [2]

(b) Write an equation to show the oxidation of ethanol to form ethanal, using [O] to represent the oxidising agent. [1]

(c) Explain why the absorption at 3400 cm^{-1} in the ethanol spectrum does not appear in the spectrum for ethanal. [1]

(d) The reaction mixture from the experiment containing ethanal was left for several days, and another infrared spectrum taken. It was observed that a new absorption was present between 2500–3000 cm^{-1}. Explain this observation. [2]

2 When cyclohexanol is heated with concentrated sulfuric acid, a reaction occurs and a new product is formed. The infrared spectrum for the new compound is shown below.

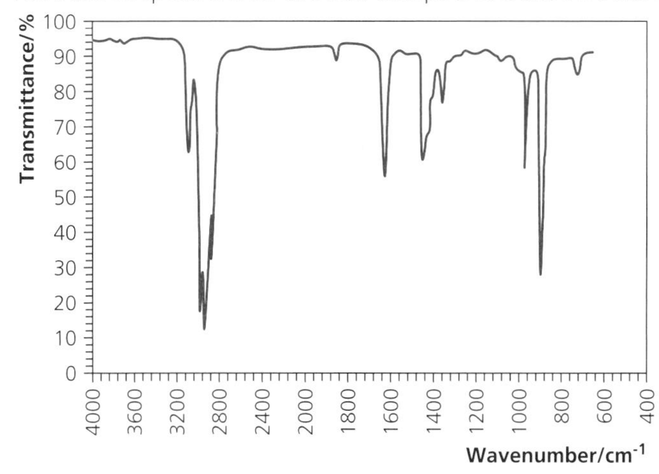

(a) Use the data in Table 15.1 (page 100) to explain how the spectrum can indicate that cyclohexanol is no longer present. [1]

(b) Using the spectrum, identify the product of the reaction giving reasons for your choice. [2]

(c) Indicate how high-resolution mass spectrometry can be used to identify the product of this reaction. [1]

Answers and quick quiz 6 online

Online

Examiners' summary

You should now have an understanding of:

- how mass spectrometry can be used to identify a substance by measuring an accurate value for the mass of its molecular ion
- how infrared spectroscopy works by measuring characteristic vibrational frequencies in molecules
- how infrared spectroscopy can be used to identify the main bonds in a molecule
- the 'fingerprint region' in the infrared spectrum and its importance when identifying a compound
- global warming in terms of molecular vibrations, the same process that takes place when measuring the infrared spectrum of a compound

Now test yourself answers

Chapter 1 Atomic structure

1 **(a)** 4 protons, 5 neutrons, 4 electrons

(b) 15 protons, 16 neutrons, 15 electrons

(c) 12 protons, 12 neutrons, 10 electrons

(d) 53 protons, 74 neutrons, 54 electrons

2 Relative atomic mass of krypton = $[(0.35 / 100) \times 78] + [(2.3 / 100) \times 80] + [(11.6 / 100) \times 82] + [(11.5 / 100) \times 83] + [(56.9 / 100) \times 84] + [(17.4 / 100) \times 86]$

= 83.93 (no units)

3 Let y = % of ^{10}B, then % of $^{11}B = 100 - y$

Therefore $[(y / 100 \times 10)] + [((100 - y) / 100) \times 11] = 10.8$

So, $0.1y + 11 - 0.11y = 10.8$

This gives: $0.2 = 0.01y$, so $y = 20\%$, so $^{10}B = 20\%$ and $^{11}B = 80\%$

4 **(a)** Lithium: $1s^2, 2s^1$; sodium: $1s^2, 2s^2, 2p^6, 3s^1$; potassium: $1s^2, 2s^2, 2p^6, 3s^2, 3p^6, 4s^1$

(b) The number of electrons in the +3 ion is 10, so the atomic number = 10 + 3 = 13

5 **(a)**

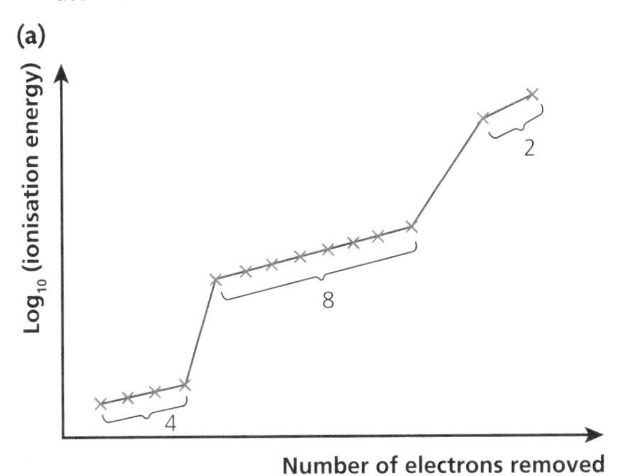

(b) $P^{3+}(g) \rightarrow P^{4+}(g) + e^-$

Chapter 2 Amount of substance

1 Amount of Ca = 1.00 / 40.1 = 0.0249 mol; amount of Cl = 1.77 / 35.5 = 0.0499 mol; ratio is 1:2; empirical formula is $CaCl_2$

2 Amount of calcium = 0.210 / 40.1 = 5.24×10^{-3}; amount of silicon = 0.147 / 28.1 = 5.23×10^{-3} mol; amount of oxygen = 0.252 / 16.0 = 0.0158 mol; ratio is 1:1:3; empirical formula is $CaSiO_3$

3 **(a)** Assuming a total mass of 100 g : amount of nitrogen = 30.4 / 14.0 = 2.17 mol; amount of oxygen = (100 − 30.4) / 16.0 = 4.35 mol; ratio is 1:2; empirical formula is NO_2

(b) Because the relative molecular mass is 92, twice the empirical formula is required; molecular formula is N_2O_4

4 Using $pV = nRT$; $p \times 0.0330 = 0.905 \times 8.31 \times (200 + 273)$; pressure is 1.08×10^5 Pa

5 Amount of Mg = 2.00 / 24.3 = 0.0823 mol; amount of hydrogen = 0.0823 mol; volume of hydrogen = nRT / p = $0.0823 \times 8.31 \times 298 / 100\,000 = 0.00204\ m^3$ or $2.04\ dm^3$

6 Ratio of nitrogen to hydrogen is 1:3; the volume of hydrogen required = $150\ cm^3 \times 3 = 450\ cm^3$

7 Amount of H_2O_2 used = (1.70 / 34.0) = 0.050 mol; amount of oxygen formed = 0.050 / 2 = 0.025 mol; volume = nRT / p = $0.025 \times 8.31 \times 298 / 100\,000 = 0.619\ dm^3$

8 Using moles dissolved = (volume (cm^3) / 1000 cm^3) × concentration (in mol dm^3):

(a) Amount dissolved = (10.0 / 1000) × 0.200 mol dm^{-3} = 2.00×10^{-3} mol NaOH

(b) Amount dissolved = (250 / 1000) × 1.20 mol dm^{-3} = 0.300 mol HNO_3

9 Amount of dissolved HCl = (20.0 / 1000) × 0.900 = 0.0180 mol; amount of NaOH reacting is 0.0180 mol; so volume reacting = 0.018 × (1000 / 0.0500) = 360 cm^3

10 Atom economy = (50.5 / (16 + 71)) × 100 = 58.0 %

11 **(a)** NaF

(b) K_2SO_4

(c) $Al(OH)_3$

12 $2Al(s) + 3CuSO_4(aq) \rightarrow Al_2(SO_4)_3(aq) + 3Cu(s)$

13 $KOH(l) + HCl(aq) \rightarrow KCl(aq) + H_2O(l)$;
$H^+(aq) + OH^-(aq) \rightarrow H_2O(l)$

Chapter 3 Bonding

1 **(a)**

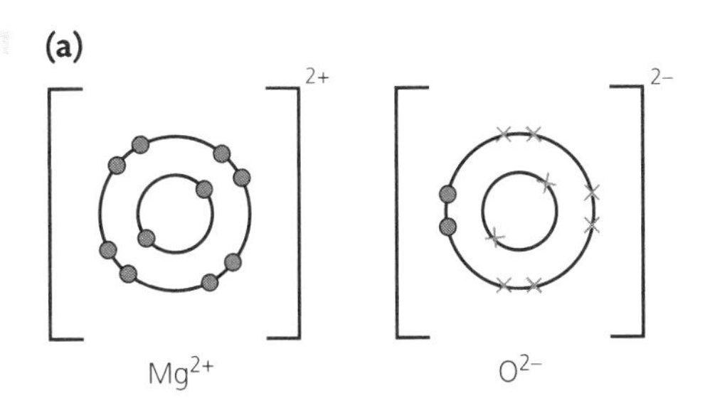

(b)

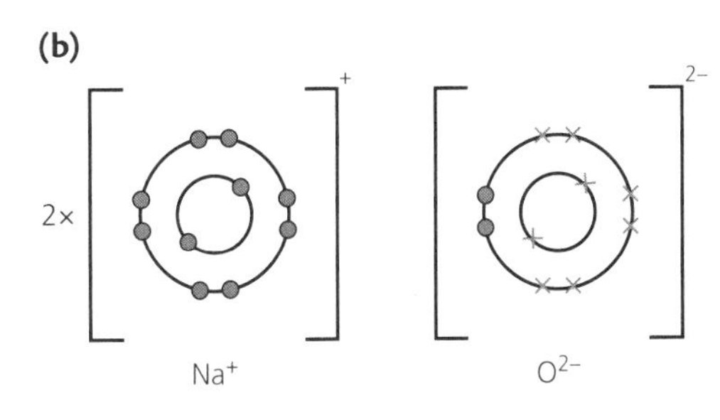

2 Lithium fluoride has a giant ionic lattice structure containing oppositely charged ions — Li^+ and F^-. These will be attracted to each other strongly by electrostatic forces; this means that they will be difficult to separate.

3 **(a)**

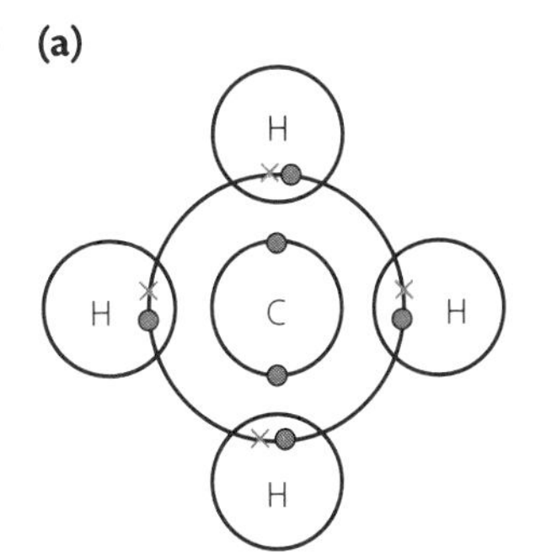

(b)

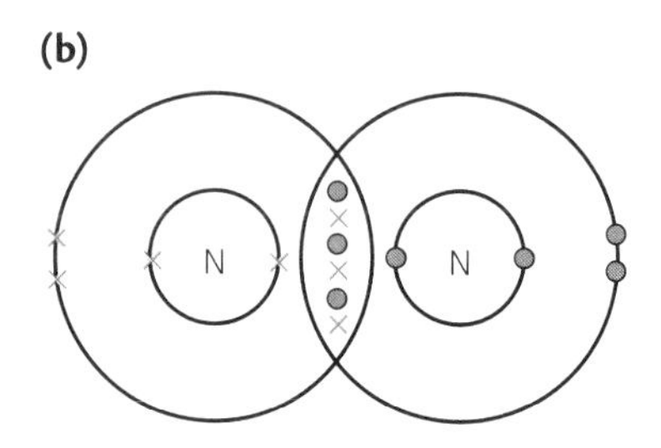

Methane is simple covalent; this means that the individual molecules will be relatively easy to separate as there are only weak forces between the molecules.

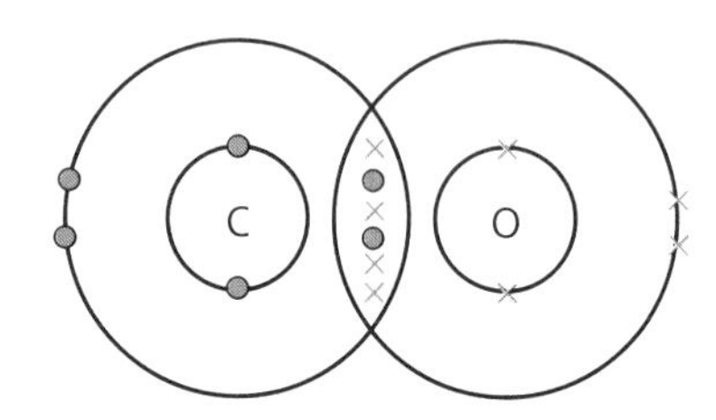

Hydrogen and fluorine have very different electronegativities; this result in a dipole forming in HF with the bonded electrons closer to the fluorine end of the molecule. In hydrogen, the atoms are identical; so their electronegativities will be the same; H_2 will therefore be a non-polar molecule.

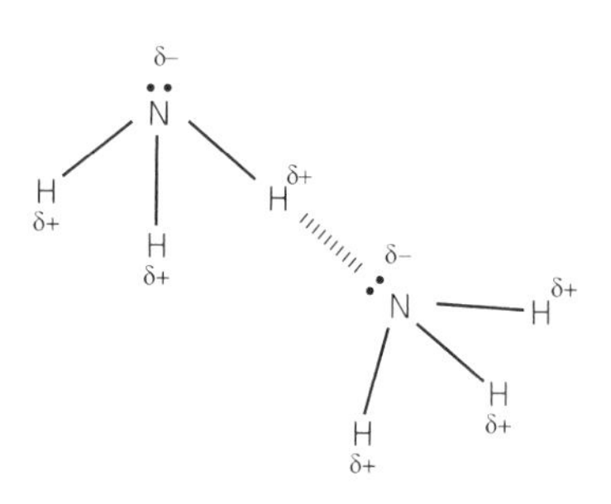

As the group is descended, molecules contain more electrons; therefore there will be greater van der Waals' forces; this means that it will be more difficult to separate the halogen molecules at the bottom of the group.

NaF, has a giant ionic lattice structure whereas F_2 is simple covalent; in NaF the oppositely charged ions will be difficult to separate because they are electrostatically attracted; in F_2 the individual molecules are easy to separate; so fluorine has a low melting point.

(a) Tetrahedral

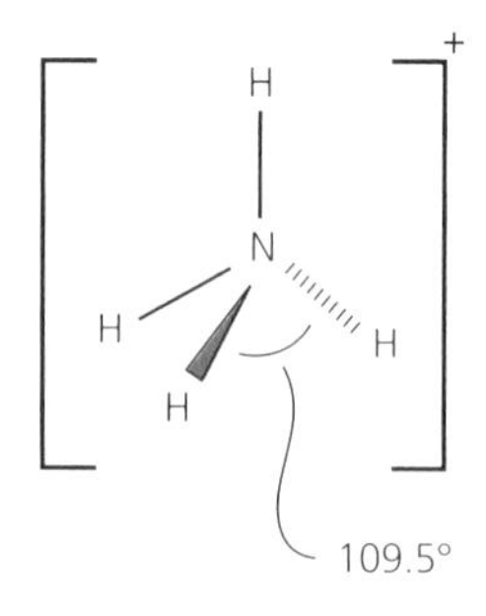

(b) Trigonal planar

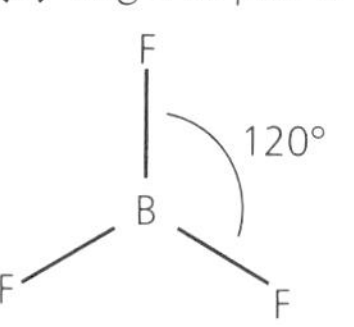

(c) Octahedral

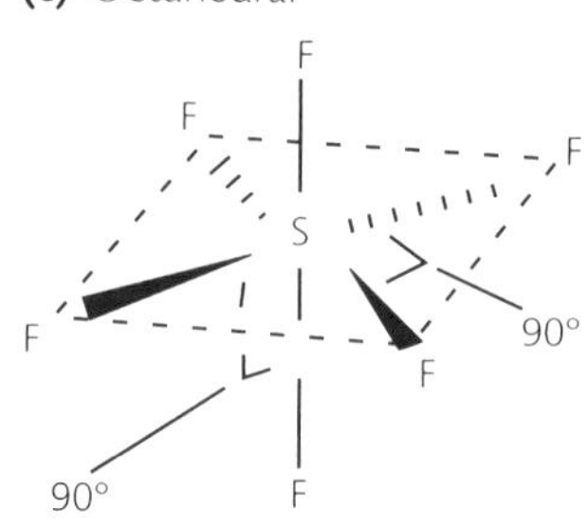

Chapter 4 Periodicity

The p block

The p block

(a) Magnesium atoms have 12 protons in the nucleus; sodium has 11; both have the same number of energy levels or shells. The outer electrons in magnesium will experience a stronger attraction; there will be a contraction in size giving a smaller atomic radius.

(b) In aluminium, the electron being removed comes from a 3p orbital which is slightly further away from the nucleus than the 3s orbital in magnesium from which its electron is removed; there is also slightly more shielding; so the electron being removed from an aluminium atom is attracted more weakly; the first ionisation energy will be lower.

(c) Both are simple covalent; but sulfur consists of S_8 molecules whereas chlorine consists of Cl_2 molecules; S_8 molecules have more electrons; and the van der Waals' forces are greater; so its boiling point will be higher.

Chapter 5 Alkanes

Displayed formula:

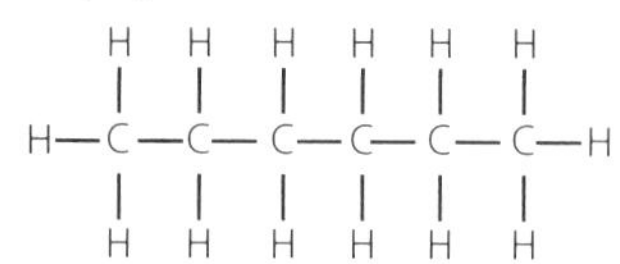

structural formula: $CH_3(CH_2)_4CH_3$ molecular formula: C_6H_{14} empirical formula, C_3H_7

Pentane, 2-methylbutane and 2,2-dimethylpropane

$C_9H_{20}(g) \rightarrow C_2H_4(g) + C_7H_{16}(g)$

$350\,cm^3$

$CH_4(g) + {}^3/_2O_2(g) \rightarrow CO(g) + 2H_2O(l)$

Oxides of nitrogen, carbon monoxide; both can be removed using a catalytic convertor

Chapter 6 Energetics

1 Using $q = mc\Delta T$; $q = 25.0 \times 4.18 \times 11.4 = 1191\,J = 1.19\,kJ$
Amount of zinc used is $0.95 / 65.4 = 0.0145$ mol; enthalpy change $= 1.19\,kJ / 0.0145\,mol = -82.1\,kJ\,mol^{-1}$

2 $q = mc\Delta T$; $q = 100 \times 4.18 \times (25.0 - 17.0) = 3344\,J = 3.34\,kJ$; 0.16 g of methanol ($CH_3OH$) is 0.16 / 32.0 mol i.e. 5.00×10^{-3} mol; enthalpy change is $-[3.34\,kJ / 5.00 \times 10^{-3}] = -668\,kJ\,mol^{-1}$

3 **(a)** $C(s) + O_2(g) \rightarrow CO_2(g)$
(b) $6C(s) + 7H_2(g) \rightarrow C_6H_{14}(l)$

4 $\Delta H = \Sigma\Delta H_f(\text{products}) - \Sigma\Delta H_f(\text{reactants} = (4 \times -1279) - ((-2984.0 + (6 \times -286)) = -5116 - (-4698.8) = -416\,kJ\,mol^{-1}$

5 ΔH_f propane $= (3 \times -393.5) + (4 \times -285.8) - (-2220.0) = -1180.5 - 1142.0 + 2220.0 = -102.5\,kJ\,mol^{-1}$

6 ΔH = energy required to break bonds – energy released on forming new bonds = $[409 + (2 \times 388) + 436)] - [163 + (4 \times 388)] = 1621 - 1715 = -94\,kJ\,mol^{-1}$

Chapter 7 Kinetics

1 As temperature increases, the molecules move with a higher kinetic energy; and a greater proportion have an energy that exceeds the activation energy; so there will be more effective collisions per unit time; and the rate increases.

2 As the concentration increases, the number of ions per unit volume also increases; there will be more collisions per unit time; and the rate of reaction increases.

3 Manganese(IV) oxide is a catalyst in the decomposition of hydrogen peroxide to form water and oxygen. The surface of the MnO_2 provides an alternative route for the reaction — one of lower activation energy.

Chapter 8 Equilibria

1 **(a)** When the temperature is increased, the equilibrium position shifts to the left-hand side by using its endothermic route.
(b) As temperature increases, the rate of reaction will always increase; more molecules have an energy that exceeds the activation energy; there will be more effective collisions taking place per unit time; and the rate will increase.

2

	Equilibrium position	**Rate at which equilibrium is attained**
Total pressure is increased	Shifts to the right-hand side	Increases
Temperature is increased	Shifts to the left-hand side	Increases
A catalyst is added	No change	Increases

Chapter 9 Redox reactions

1 **(a)** +3; **(b)** +1; **(c)** +4; **(d)** +6; **(e)** +6; **(f)** +5; **(g)** +6

2 **(a)** cobalt(III) chloride; **(b)** sodium chlorate(I);
(c) titanium(IV) chloride **(d)** sodium ferrate(VI);
(e) sulfuric(VI) acid; **(f)** iodate(V) ion **(g)** manganate(VI) ion

3 Oxidising agent: oxygen gas, O_2; reducing agent: SO_2

4 Pb: +4 to +2; Pb(IV) is the oxidising agent; Cl: −1 to 0; Cl^- is the reducing agent. (Some Cl^- is unchanged)

5 **(a)** $Zn(s) \rightarrow Zn^{2+}(aq) + 2e^-$; $Fe^{2+}(aq) + 2e^- \rightarrow Fe(s)$
(b) $2Al(s) \rightarrow 2Al^{3+}(aq) + 6e^-$; $3Cu^{2+}(aq) + 6e^- \rightarrow 3Cu(s)$

Chapter 10 The periodic table

1 Chlorine has fewer electron energy levels compared to iodine; so the added electron will be closer to the nucleus in the chlorine atom; and the electrostatic attraction will be greater; so chlorine will be the more powerful oxidising agent.

2 **(a)** $Br_2(aq) + 2NaI(aq) \rightarrow 2NaBr(aq) + I_2$
or $Br_2(aq) + 2I^-(aq) \rightarrow 2Br^-(aq) + I_2(aq)$
(b) Bromine has been reduced; iodide ions have been oxidised.

3 $KF(s) + H_2SO_4(l) \rightarrow KHSO_4(s) + HF(g)$

4 **(a)** H_2S: −2; SO_2: +4; S: 0
(b) $6HI(g) + H_2SO_4(l) \rightarrow 3I_2(s) + S(s) + 4H_2O(l)$

5 To a solution of each salt; add dilute nitric acid and then silver nitrate solution; a white precipitate of AgCl forms with lithium chloride; and a yellow precipitate of AgI forms with lithium iodide:
$Ag^+(aq) + Cl^-(aq) \rightarrow AgCl(s)$
$Ag^+(aq) + I^-(aq) \rightarrow AgI(s)$

6 **(a)** Chlorine: purifying drinking water; sodium chlorate(I) : bleach
(b) Water and chlorine gas

7 $Ca(s) + 2H_2O(l) \rightarrow Ca(OH)_2(aq) + H_2(g)$
Bubbles would be observed; calcium metal (a light-grey solid) would eventually dissolve to form a white suspension in water.

8 Add barium chloride solution and hydrochloric acid to separate solutions of sodium sulfate and sodium nitrate; a white precipitate forms with sodium sulfate; there is no reaction with sodium nitrate.

Chapter 11 Extraction of metals

1 **(a)** **(i)** $2O^{2-}(l) \rightarrow O_2(g) + 4e^-$
(ii) $Al^{3+}(l) + 3e^- \rightarrow Al(l)$
(b) Molten cryolite mixes with aluminium oxide; to form a mixture with a lower melting point; it therefore reduces energy costs.

2 Creates less waste; conserves resources such as metal ores; reduces energy costs; reduces air pollution.

Chapter 12 Haloalkanes

(a) CFCs produce chlorine radicals in the presence of ultraviolet light; these then attack ozone molecules

(b) A particle (atom, molecule or ion); with one or more unpaired electrons

(c) $Cl\bullet + O_3 \rightarrow ClO\bullet + O_2$ then $ClO\bullet + O_3 \rightarrow 2O_2 + Cl\bullet$

(a) $C_2H_6(g) + Br_2(g) \rightarrow C_2H_5Br(g) + HBr(g)$

(b) Free-radical substitution

(c) Initiation: $Br_2 \rightarrow 2Br\bullet$

Propagation: chlorine radicals react with methane molecules to form new radicals and molecules:

$C_2H_6 + Br\bullet \rightarrow HBr + \bullet C_2H_5$
then $\bullet C_2H_5 + Br_2 \rightarrow C_2H_5Br + Br\bullet$

Termination: radicals react with each other to form molecules:

$\bullet C_2H_5 + Br\bullet \rightarrow C_2H_5Br$ or $\bullet C_2H_5 + \bullet C_2H_5 \rightarrow C_4H_{10}$

Chapter 13 Alkenes

```
     H                H   H
     |                |   |
   H-C                C - C-H
    / \             / |   |
   H   C = C          H   H
      /     \
     H       H
```

Z-pent-2-ene

```
   H   H
    \ /        H
     C        /
    / \      /
   H   C = C     H   H
      /     \    |   |
     H       C - C-H
             |   |
             H   H
```

E-pent-2-ene

(a)

```
H         CH3
  \      /
   C = C
  /      \
H         CH3
```

(b)

```
 H
  \        CH3                        CH3
   \    + /                  +       /
H - C - C           or   H - C - C - CH3
   /      \                  |   |
 H         CH3               H   H
```

(c) The tertiary carbocation:

```
       +
H3C -  C - CH3
       |
       CH3
```

(d)

```
       Br
       |
H3C -  C - CH3
       |
       CH3
```

(a)

```
    H   H   H
    |   |   |
H - C - C - C - H
    |   |   |
    H   Br  H
```

(b)

```
    H   H   H
    |   |   |
H - C - C - C - H
    |   |   |
    Br  Br  H
```

(c)

```
    H   H   H
    |   |   |
H - C - C - C - H
    |   |   |
    H   O   H
        |
      O=S=O
        |
        OH
```

(d)

```
    H   H   H
    |   |   |
H - C - C - C - H
    |   |   |
    H   OH  H
```

```
  CH3 H  CH3 H  CH3 H
   |  |   |  |   |  |
 - C -C - C -C - C -C -
   |  |   |  |   |  |
   H  H   H  H   H  H
```

Chapter 14 Alcohols

```
    H   H   H   H   H
    |   |   |   |   |
H - C - C - C - C - C - O - H
    |   |   |   |   |
    H   H   H   H   H
```

```
    H   H   H   H                    H   H   H   H
    |   |   |   |     //O            |   |   |   |     //O
H - C - C - C - C - C            H - C - C - C - C - C
    |   |   |   |     \              |   |   |   |     \
    H   H   H   H      H             H   H   H   H      O - H
```

pentanal

pentanoic acid

```
    H   O   H   H                   H   H   H
    |   ||  |   |                   |   |   |     //O
H - C - C - C - C - H           H - C - C - C - C
    |       |   |                   |   |   |     \
    H       H   H                   H   H   H      H
```

butanone

butanal

Silver nitrate and ammonia solution (Tollens' reagent); is heated separately with both substances; butanal will form a silver mirror; butanone will not.

Chapter 15 Analytical techniques

1

```
     H   H   H
     |   |   |
H —  C — C — C — O — H
     |   |   |
     H   H   H
```

2 C–C; C–H; C–O; O–H

3 Likely wavenumbers: 3000–3300 cm^{-1}(broad) for the O–H; 1000–1300 cm^{-1} for the C–O

4 Likely to be due to a C=O bond; the sample therefore contains an impurity such propanal, or propanoic acid; formed by the oxidation of propan-1-ol.

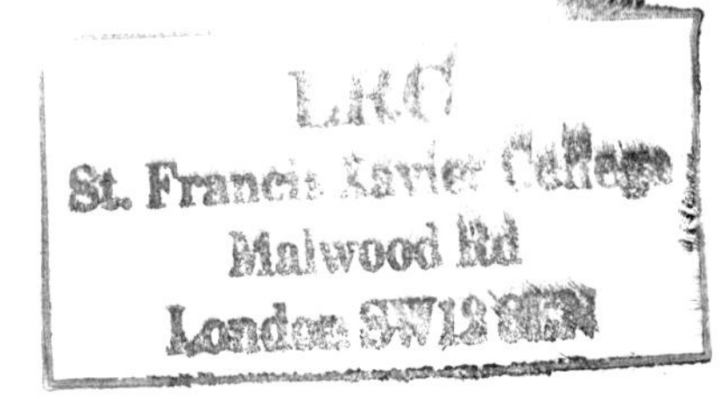
LRC
St. Francis Xavier College
Malwood Rd
London SW12 8EN